OWENS VALLEY REVISITED

OWENS VALLEY REVISITED

A Reassessment of the West's First Great Water Transfer

GARY D. LIBECAP

STANFORD ECONOMICS AND FINANCE
An Imprint of Stanford University Press
Stanford, California

Stanford University Press
Stanford, California

Printed in the United States of America on acid-free, archival-quality paper

Library of Congress Cataloging-in-Publication Data

Libecap, Gary D.
    Owens Valley revisited : a reassessment of the West's first great water transfer / Gary D. Libecap.
        p.   cm.
    Includes bibliographical references and index.
    ISBN 978-0-8047-5379-1 (cloth : alk. paper)—ISBN 978-0-8047-5380-7 (pbk. : alk. paper)
        1. Water transfer—California—Owens Valley—History.   2. Water transfer—Law and legislation—West (U.S.)   3. Water-supply—California—Los Angeles.   4. Owens River Watershed (Calif.)—Water rights—History.   5. Water rights—West (U.S.)   I. Title.

F868.O9L53   2007
363.6'10979494—dc22   2007001773

Typeset by G&S Book Services in 10/13.5 Minion

Special discounts for bulk quantities of Stanford Economics and Finance books are available to corporations, professional associations, and other organizations. For details and discount information, contact the special sales department of Stanford University Press. Tel: (650) 736-1783, Fax: (650) 736-1784

FOR ANN, SARAH, AND CAP, WHO LOVE WILD WATER AS MUCH AS I DO

CONTENTS

ACKNOWLEDGMENTS

Support for this research was provided by National Science Foundation Grant 0317375, co-principal investigators Alan Ker and Robert Glennon; the Robert Wesson Fellowship at the Hoover Institution, Stanford University; the Julian Simon Fellowship at the Property and Environment Research Center (PERC), Bozeman, Montana; the Earhart Foundation; the International Center for Economic Research (ICER), Turin, Italy; and the Anheuser-Busch Chair, McGuire Center for Entrepreneurship, Eller College, University of Arizona.

OWENS VALLEY REVISITED

A REEXAMINATION OF OWENS VALLEY AND WESTERN WATER

I said, "What was the fight over? " and Mr. Tripp said "Same old thing—water."

E. F. Leahey, Land Agent, Los Angeles Department of Water and Power

This book reexamines the infamous Owens Valley–to–Los Angeles water transfer and the negotiations between valley farmers and representatives of the Los Angeles Department of Water and Power (LADWP) that took place from 1905 through 1935 over the sale of land and water rights. After that, at times Owens Valley supplied nearly 80 percent of Los Angeles's water, although recently that figure has dropped to around 34 percent.[1] This water exchange, now over seventy years old, still plays an important role in perceptions of how water markets work. The book also examines the experience of Los Angeles's water rights in Owens Valley and the Mono Basin since 1970. New environmental and recreational values and related physical and pecuniary externalities brought demands to curtail the shipment of water from the source regions for urban use. How those demands were addressed, the likely costs involved, and alternative approaches that might have resulted in more rapid and less contentious remedies are explored.

THE LESSONS OF OWENS VALLEY

The allocative benefits of moving water from historical, low-value uses in agriculture to new, high-value ones often are so great on the margin that considerable surplus is generated. How those surpluses are distributed affects the positions of the parties involved in water exchanges and how they interpret the desirability of water markets. This was the situation in Owens Valley in the 1920s and it remains so today, elsewhere in the American West. Although unlike Owens Valley most contemporary agriculture-to-urban water transfers do not involve more than a fraction of a region's water and thereby have much less total impact on the local economy and society, the lessons of Owens Valley remain pertinent for today's policy and academic discussions about water transfers.

Unfortunately, the legacy of Owens Valley for water markets is a decidedly negative one. Whenever farmers are asked about the sale of some of their water rights to cities, invariably the response from some is that they "do not want to be another Owens Valley." Many of the given "lessons" of Owens Valley for today's water problems, however, are incorrect. They have been distorted by over a half-century of telling and retelling the Owens Valley story in order to meet the varying agendas of the tellers. Along the way, many of the facts of the exchange have been lost, and some of the history of the land and water exchange has never been thoroughly examined. Owens Valley does offer lessons for contemporary water trades, but they are different from prevailing myths and those that have been asserted in the literature. And these lessons are more useful for understanding the problem of water and its allocation and for promoting more socially valuable use of the West's most precious resource.

Perhaps the most important lesson of Owens Valley is how intertwined the distributional and efficiency effects of water reallocation are. Efficiency typically drives the incentives to transfer water from low- to higher-value uses. But because of a multitude of interrelated, simultaneous, or sequential uses of water, a seemingly simple transfer between two parties can become complicated because of the impact on third parties.

Buyers and sellers naturally compete over the benefits of any trade. In Owens Valley, most of the disputes were between Los Angeles and farmers who owned the land and water rights. The bargaining strategies used by each had important effects on the transaction costs of exchange and the overall perception of fairness. The direct payments to water rights holders, however, are much less difficult to determine than is compensation for third-party effects.[2]

In Owens Valley there were disputes between representatives of the city of Los Angeles and town property owners who charged that they were harmed by the falloff in commercial activity due to the sale of water. Resolving third-party effects such as these involves the thorny problems of assessing legitimacy, measurement, payment, and limiting claims to control rent-seeking.[3]

It may be that the efficiency gains of water transfer swamp these distributional concerns so that, on net, society is better off from the transfer. Nevertheless, unless effectively addressed, the political and judicial reaction to third-party grievances can block, severely limit, or slow otherwise socially beneficial water transfers.[4]

Water is a particularly sensitive resource in the semi-arid West. Indeed, unlike land, water is owned by the state, with individuals having only use rights. Its allocation and application receive considerable scrutiny, and there is a long history of conflict over it. As Mark Twain is famously quoted, "Whiskey is for drinking; water is for fighting over."[5] The legacy of Owens Valley and the negotiations over its water rights are discussed in Chapters Two through Five.

Another lesson of Owens Valley is the need to compensate water rights holders for new environmental and recreational restrictions on their ability to use or transfer water they obtained in the past. As we will see, in the 1920s and 1930s when Los Angeles obtained most of its water rights in Owens Valley and the Mono Basin, urban development was the highest-value use of water. At the time, few questioned the ultimate destination for the water or were concerned about any physical effects of the transfer. Most of these were, in any event, unanticipated, so it was not possible to specify ex ante how much water could be withdrawn from the region without harming the environment. Rather, the controversy of the day was over the division of the surplus.

In the 1970s, however, when the city sought to export more of its water from the region to meet growing urban demand, objections were raised that doing so was damaging the environment and reducing recreational opportunities. Environmental and local groups demanded that water diversions be reduced. Because Los Angeles held the water rights, there was a basis for exchange to secure water to mitigate any physical effects of its export. Market transactions may have been feasible, at least in some cases. If concerns over the city's position as a single seller or the magnitude of the transaction costs associated with private restoration projects ruled out direct use of the market, eminent domain with compensation might have been used by government agencies to acquire water.

Instead, lengthy and divisive court battles as well as administrative reviews ensued. The adversarial, all-or-nothing nature of the litigation seems to have both delayed the response to local environmental concerns and raised the transaction costs of addressing them. Moreover, it is unlikely that the process generated the information necessary to determine the optimal amount of water to be retained in the watershed or to be exported for urban use. These issues are addressed in Chapters Six through Eight.

A final lesson of Owens Valley is the special problem of groundwater withdrawal and management. The hidden and fluid nature of groundwater basins makes the effective allocation of water rights more difficult. Bounding allotments and monitoring extraction are very hard to do. There is always guesswork in determining how much pumping can occur before causing injury to neighboring rights holders. Many western groundwater basins are effectively common-pool resources with the potential to create a "tragedy of the commons" due to physical externalities from competitive drawdown.[6]

In Owens Valley in the 1920s, Los Angeles acquired farm properties not only to secure their surface and groundwater water rights, but also to internalize the externalities of groundwater pumping. In part, the city was forced to do so because farmers who could demonstrate subsidence or other damages from the operation of city pumps could obtain a court order to halt all pumping in the basin. Because the city relied on pumping as the residual source of supply of water for the Los Angeles Aqueduct, it could not allow pumping to stop. Accordingly, Los Angeles strategically purchased farms whose owners threatened legal action to halt pumping even if they otherwise had little water for the aqueduct. Ultimately, the city acquired virtually all of Owens Valley. In this way, Los Angeles came very close to fully "unitizing" or internalizing the negative physical effects of pumping.[7] Because surface ownership is so fragmented, this outcome rarely occurs above other major aquifers today, so some form of basin-wide regulation of withdrawal often is warranted.[8]

Ironically, this effort did not protect Los Angeles from subsequent demands to regulate city pumping. Although there were criticisms from ranchers who leased city lands in the 1930s when persistent drought brought greater groundwater withdrawals, the major reaction occurred after 1970, when more extensive extraction was used to help fill the recently completed second Los Angeles Aqueduct. Complaints were directed at externalities that were less completely internalized—notably dust pollution off the dry Owens Lakebed. Complaints also focused on the deterioration of aquatic and riparian habitat as surface water sources in communication with groundwater dried up. In this controversy, there were problems of measurement and of separating physical and pecuniary effects, as well as of determining the most effective regulatory response.

NEW RESEARCH ON OWENS VALLEY

Despite the notorious history of Owens Valley and the allegations of water theft and plundering by Los Angeles associated with it, there has been no systematic analysis of the underlying negotiations between the Board of Water and Power Commissioners and farmers, nor has there been careful examination of what happened to the economy and population of the valley once ownership passed to the LADWP. Both of these analyses are provided here. We will see why the bargaining between the agency and landowners was so acrimonious for some, whereas for others there were no disputes and both parties were satisfied with the outcome.

The analysis reveals what the farmers received for their lands and their water, how those sums varied according to farm characteristics and seller pool membership, and how they compared with what the LADWP might have been willing to pay for water. The impacts of the land and water purchases on Owens Valley farmland prices, production, and farm size, as well as the effects on the population and assessed property values of Inyo County (Owens Valley) and its county seat (Bishop) are presented and compared to conditions in similar counties elsewhere in California and Nevada. All told, Owens Valley landowners did well in their land sales, earning more than if they had stayed in agriculture, and the valley was not plundered. The rural population of the county did not collapse, nor did the agricultural economy disappear. This is a much more positive outcome than is portrayed in contemporary accounts of Owens Valley.

But most farmers received far less for their water than Los Angeles might have been willing to pay. Battles over the distribution of the large gains from reallocating Owens Valley water were the source of the conflict between some farmers (usually those with the most water) and the LADWP. Indeed, had farmers been able to capture more of those returns, we might hear less about Owens Valley now.

Before turning to both the history and recent experience of water rights in Owens Valley, it is important to place the Owens Valley transfer, the first major agriculture-to-urban water exchange, into the context of contemporary water markets. There is a considerable gap between water values at the margin in urban and environmental uses and historical use values in agriculture, where 80 percent of consumptive use takes place. And this gap is growing. The process of water transfers and water marketing remains controversial for many of the same issues encountered in Owens Valley.[9] To understand why, the nature of western water rights, the interconnected private and public uses of water, and state regulation of water exchanges are briefly summarized.

THE MOTIVATION FOR WATER TRANSFERS: RISING URBAN DEMAND

As was the case with Owens Valley, there are contemporary social gains from moving water from agriculture to meet urban demand. In 1992, Griffin and Boadu reported that the value of water used in agriculture, capitalized over

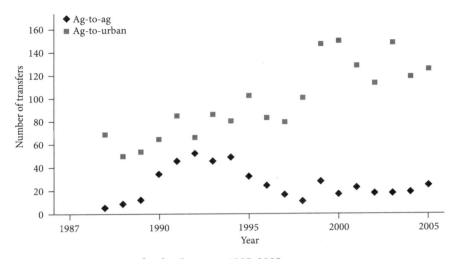

Figure 1.2 Water Transfers by Category, 1987–2005.

SOURCE: Glennon, Ker, and Libecap (2007), using data from *Water Strategist.*

Figure 1.1. Indeed, much of the observed increase in agriculture-to-urban trans-fers comes from activity in the Colorado–Big Thompson Project/Northern Colo-rado Water Conservancy District, which has more refined water rights and trans-fer processes than are found in most other parts of the West. If Colorado trades are removed from the data, there still is a shift toward agriculture-to-urban transactions, but the pattern is not as pronounced as indicated in the figure.[15]

In fact, in the West, most of the water that is traded remains within sectors. While agriculture-to-urban transfers of the kind that took place in Owens Valley constituted 56 percent of the number of water trades that took place from 1987 through 2005, they included just 18 percent of the total water moved.[16] In con-trast, agriculture-to-agriculture trades, 15 percent of the total number, involved 23 percent of the volume. Agriculture-to-agriculture and urban-to-urban trades jointly included 41 percent of all water exchanged over the nineteen years.[17] In addition, within-sector trades were often negotiated as one-year leases, whereas 86 percent of agriculture-to-urban exchanges were permanent sales.[18] Within-sector trades are much less controversial and have lower transaction costs than agriculture-to-urban transfers.[19]

By keeping the water in the source area, within-sector trades minimize third-party effects and the regulatory response to them. Conveyance costs also are generally lower. Finally, within-sector trades, especially in agriculture, can in-volve water of relatively low quality and low-priority appropriative water rights.

Because agriculture-to-agriculture trades are usually through one-year leases, issues of the long-term security of the water right, as may arise in exports of water to urban use, are not encountered. And agricultural applications do not require the quality of water necessary for urban consumption.

The largest gains from trade, however, are with agriculture-to-urban transfers, and as with Owens Valley, contemporary water trades require property rights. The following sections explore the nature of property rights to water and regulation of water transfers to address third-party effects.

WESTERN WATER RIGHTS

Many of the problems of trading western water lie in the complex system of property rights and the difficulty of defining them. In western states, water is "owned" by the state in trust for its citizens and its use regulated, based on public interest or welfare concepts.[20] As stated in Wyoming law, for example: "Because water is so important to the economy of this state, its use is always limited by a concept of public trust; the only uses for which water rights may be established are those which receive 'public recognition' under the law of the state."[21]

Individuals hold usufruct rights to the water, subject to the requirement that the use be beneficial and reasonable and to oversight by the state in monitoring transfers.[22] In some cases, as occurred in the Mono Basin, there is also application of the public trust doctrine to limit private use, diversion, and restrictions on access.[23] For these reasons, western water rights appear to have less protection or to be more fragile than most other property rights to tangible resources.[24] In Owens Valley, farmers held both appropriative and riparian surface water rights and groundwater rights.

Appropriative Surface Water Rights

Most western water rights are based on prior appropriation. The appropriative doctrine allows rights holders to withdraw a certain amount of water from its natural course for private beneficial purposes on land remote from the point of diversion.[25] The maintenance of appropriative rights is based on placing claimed water into beneficial use. Because beneficial uses are difficult to measure, the basic test of meeting the beneficial use requirement has been physical diversion. The twin requirements of diversion and beneficial use reflect the long-standing emphasis on using scarce water for economic growth. Because appropriative rights can be separated from the land and sold or leased, they are the basis for water transfers in response to changing economic conditions.[26]

The appropriative doctrine emerged in the nineteenth century in response to the development of mining and agriculture in the semi-arid West, where growing numbers of people and economic activities were increasingly concentrated in areas where there was too little water.[27] The most arid western states—Arizona, Colorado, Idaho, Montana, Nevada, New Mexico, Utah, and Wyoming—constitutionally or statutorily adopted the appropriative system.[28]

Ownership of water is assigned through the rule of first possession or priority of claim.[29] Those with the earliest water claims have the highest priority, and those with subsequent claims have lower priority or junior claims. There is a ladder of rights on a stream, ranging from lowest in priority to highest. This allocative mechanism provides a clear way of ranking competing claimants in assigning rights and in rationing water during times of drought.

Riparian Surface Water Rights

Riparian rights are common-law institutions that dominate in the eastern United States, but they also are recognized, along with appropriative water rights, in parts of California, Nebraska, Oklahoma, Oregon, Washington, North and South Dakota, and Texas.[30] Even so, the appropriative system is used most widely. In the East and in western states with more precipitation, there is less need to divert and ship water long distances to more arid sites.

Riparian rights are based on ownership of land appurtenant to water flows. These landowners have rights to access the water adjacent to or passing through their properties for reasonable use, including fishing and navigation, and they can use the water so long as doing so does not harm other riparian claimants downstream.[31] Riparian water rights can only be transferred along with adjacent property, and they are not lost through disuse.

Groundwater Rights

Groundwater rights vary across the western states; most are not well defined or enforced.[32] As with surface water rights, prior appropriation and reasonable use are the dominant allocative mechanisms.[33] Seniority entitles surface owners the right to extract a "reasonable amount" of water from below their property, however that quantity is defined.[34]

Depending on subsurface pressure and permeability of the soil, water may migrate as pumping or other forms of extraction occur. When it does, it moves through the aquifer, potentially draining the water available to others, depleting surface flows, harming associated aquatic and riparian habitats, and causing land subsidence.[35] Although groundwater is recognized as a "public resource" in western states, there has been little limitation on competitive groundwater withdrawal, and no clear recognition in the water rights structure of interconnectedness between surface and groundwater uses.[36] The requirements that a user document damage caused by the extraction of groundwater by another user are significant, raising the costs of individual attempts to enforce water rights.

State Regulation of Water Rights Transfers

Because it is difficult to segment water into its various concurrent or sequential uses, there often is a high degree of interaction among water claimants and applications.[37] Accordingly, any trades that change the location of water diversion, nature of use, and timing, especially if they are large relative to stream flow, are restricted by state law and regulated by state agencies.[38] State water agencies typically allow changes in diversion and location for only historical consumptive uses.[39] Moreover, because of the potential for harm, transfers of surface water rights in western states are predicated on there being "no harm or injury" to downstream rights holders.[40] In contrast, local water transfers within sectors typically do not require state approval.[41]

CONCLUSION

This chapter has summarized some of the lessons of Owens Valley and placed them in the context of water transfers in the American West. The conflict between Owens farmers and representatives of the Los Angeles Department of Water and Power in the 1920s and 1930s would be of only historical interest, except that many of the same problems encountered then are found in today's water exchanges. With this background on the state of water rights and water trades, we now turn to the legacy of Owens Valley and how it has shaped contemporary perceptions about the efficacy of water transfers from agriculture to urban and environmental uses.

2 | THE OWENS VALLEY SYNDROME

And at last the drop that fell as a snowflake upon the Sierra's crest and set
out to find its home in the sea, shall be taken up from beneath the ground
by a thirsty rootlet and distilled into the perfume of an orange blossom in a
garden of the City of the Queen of the Angels.

Complete Report on Construction of the Los Angeles Aqueduct, 1916

Dry Ditches
In a bleaching land,
A broken pane,
A swinging door,
And out upon
A withered field,
Where blue blossoms
Once nodded in the sun,
A rusted plow,
Deep furrow
In the crusted sand.

*Richard Coke Wood, The Owens Valley and the Los Angeles Water
Controversy*, Owens Valley as I Knew It, *1973, 1*

The quotations above represent the controversial legacy of Owens Valley. While
its water provided for the growth of Los Angeles, which is the home of millions of
people, this outcome is not widely viewed as positive. Rather, the overwhelming
view of the water transfer in the academic and popular presses is a very negative
one. The general themes are theft of the valley's water; destruction of the local,
agricultural economy; and colonization (hydrocolonialism) of the region by a remote, disinterested city.[1] According to the popular view, powerful, unscrupulous
promoters of Los Angeles grabbed the valley's water and "flushed" it down the
Los Angeles Aqueduct, converting previously verdant farmland of orchards and
other irrigated crops into desert and leaving the skeletons of abandoned home-
steads, empty schoolhouses, and dry ditches. This is a powerful image that looms
as a parable, the "Ghost of Owens Valley," affecting contemporary rural-to-urban
water transfer negotiations.[2] Such an outcome, of course, is not the future members of rural communities want as they consider possible sales of water today.

A summary of interpretations of the Owens Valley water transfer to Los Angeles as it is portrayed in the recent academic and popular presses provided in
this chapter shows how it has shaped the way in which rural-to-urban water exchanges are viewed. A leading historian of Owens Valley, William Kahrl, whose

work has appeared both as a monograph and in a historical journal, later reprinted in a law review, provides one of the most commonly referenced accounts
of water transfer.[3]

Kahrl begins by writing glowingly of the valley's agricultural potential in the
early twentieth century, which he compares favorably with that of the Imperial
Valley and was more attractive at the time than that of either the Sacramento or
the San Joaquin Valley.[4] This promise, however, was derailed when former Los
Angeles mayor Fred Eaton, Chief Water Engineer William Mulholland, and J. B.
Lippincott, chief of Southwest Operations for the Federal Reclamation Service
conspired to block a proposed Bureau of Reclamation project in 1904 that would
have provided needed drainage for the valley and to secure a right-of-way for the
aqueduct across federal land.[5] A water supply crisis was manufactured by Mulholland in Los Angeles in order to secure a favorable vote by the city's citizens
on the bonds needed to finance construction of the aqueduct. The bonds were
passed, and excavation began in 1907 with completion of the aqueduct in 1913.[6]

Once the aqueduct was in place, the fate of the valley was sealed. Representatives of the city secretly bought up land and water rights from unsuspecting
farmers.[7] Local residents were ill-prepared to meet the hardball negotiating tac-
tics of city officials, which included undermining an important local bank in August 1927 and along with it, farm mortgages, local savings, and the will to resist
the unceasing demand for more land and water.[8] Eventually, virtually all of the
valley's agricultural land was handed over to Los Angeles, and irrigated agriculture ended as the water was diverted to the aqueduct. The farm economy with all
of its potential was crushed, leaving the region dependent on the whims of the
Los Angeles Department of Water and Power (LADWP) as its colonial master.[9]
To add further insult to the valley, more than half of the early water drained
from it went not to urban residents of Los Angeles, but instead to farmland in
the San Fernando Valley, fueling land speculation there.[10]

Kahrl concludes: "And so, with money, guns, and a unity of purpose with
what they identified as the public interest, the bankers and businessmen of Los
Angeles determined to seize the water resources of Owens Valley 240 miles to
the northeast. And, by correcting God's design for their community with the
construction of the Los Angeles Aqueduct, they laid the foundations for the
modern metropolis."[11]

Various aspects of this bleak story are incorporated by legal scholars in articles about water transfers. David Getches characterizes "L.A.'s raid on the Owens Valley water" as "legendary."[12] David Howard Davies points to the allegedly
secretive way that Los Angeles bought up the farms and ranches in the valley so

that the local people could not organize opposition.[13] Robirda Lyon comments on the demise of Owens Valley, both economically and ecologically, as it lost its water.[14] Hal K. Rothman claims that Owens Valley residents lost their culture in a type of "social genocide" that rendered them "not only irrelevant but destitute."[15] And Andrew H. Sawyer described the dashed expectations of farmers as the LADWP forced them to sell through its monopsony purchasing practices.[16]

The popular press, perhaps not surprisingly, is no more charitable in its description of Owens Valley. Although this press includes numerous books, newspapers, and magazine articles, only a few representative examples are cited here. More illustrations are provided later in the chapter. As one example, a 1985 article in the *Washington Post* emphasized the desire of Imperial Irrigation District farmers to avoid an "Owens Valley" outcome in their negotiations with the Metropolitan Water District of Southern California over the sale of water to cities: "Growers here remember the Owens Valley, an oasis in the Sierras until a rapacious water grab by Los Angeles turned it into a parched desert nearly 50 years ago."[17] The perception of land theft and community destruction also was stressed almost twenty years later in a 2004 article in the *New York Times* titled, "Los Angeles Mayor Seeks to Freeze Valley Growth. Century-Old Land Grab Still Contentious."[18] In an article on Owens Valley in the *Smithsonian Magazine*, Mark Wheeler claimed that Los Angeles's acquisition practices "may fairly be called ruthless."[19] Of course, the 1974 movie *Chinatown*, starring Jack Nicholson and Faye Dunaway, added to the popular notoriety of Owens Valley by dramatizing alleged conspiracies involving the valley's water and land speculation in Los Angeles.

This gloomy and indeed sinister portrayal of the Owens Valley experience was harmful for the subsequent development of water markets not only in California, but throughout the West. Academic researchers and policy experts, such as Vincent Ostrom; Brent Haddad; and Brian Gray, Bruce Driver, and Richard Wahl, cited an "Owens Valley syndrome" as deterring efforts to reallocate water from agriculture to urban centers.[20] Ostrom claimed it deterred efforts to reallocate water from northern California to urban centers in the south, and Haddad argued that the "Ghost of Owens Valley" inhibited *all* proposed water transfers from rural areas to cities in the West. Ellen Hanak stated that "Owens Valley has indelibly marked the regional consciousness on water issues." Further, Ellen Hanak and Caitlin Dyckman, Robert Beck, and A. Dan Tarlock credited Owens Valley in motivating county-of-origin restrictions on water transfers in California.[21] Pointing to the example of Owens Valley, Aaron Ralph asked whether state regulators should allow one community to use another's water in order to grow and leave the other behind.[22] Concerns about potential negative economic effects on the exporting region also were stressed by Barton Thompson.[23] Other condemnations of the Owens Valley transfer are found in works by Richard Coke Wood, Marc Reisner, and Rebecca Ewan.[24] More balanced assessments are found in works by Remi Nadeau, Abraham Hoffman, Peter Vorster, John Walton, and Robert Sauder.[25]

The story of Owens Valley, then, is of more than historical interest. It exerts an important drag on discussions of agriculture-to-urban water transfers. Given all of this, it is worthwhile to review the legacy of Owens Valley as it is portrayed in the popular and academic literatures, in this chapter, and then in the following chapters, to reexamine the story to better understand the relationships between farmers and the LADWP, the outcomes of the land and water rights purchases, and the sources of the disputes that took place in the early twentieth century.

THE "OWENS VALLEY SYNDROME" IN THE ACADEMIC LEGAL LITERATURE

Although economists and political scientists are actively involved in research on water markets, they have not examined Owens Valley in any depth. Although they have recognized its negative impact on contemporary water transfers, the most extensive discussions of Owens Valley have been by legal scholars, concerned with both the process and the outcome of large water exports from rural areas. The way in which the Owens Valley water transfer is interpreted in law journals, however, is of more than academic interest because it directly influences how practicing lawyers, legislators, agency administrators, and judges view the reallocation of water in the West. And the picture they are provided is not a pretty one.

For example, Aaron Ralph emphasizes the potentially ruinous impact of water exports on the local economy of the exporting region. After noting that the Owens Valley–to–Los Angeles water transfer was the first major agriculture-to-urban water exchange in California, Ralph states: "Although the Owens Valley/Los Angeles water transfer won approval almost 100 years ago, it remains a controversial issue today. In fact, the devastating effects of this transfer on the Owens Valley provided the impetus for California to enact laws restricting water transfers."[26] The devastation alluded to by Ralph is not based on new empirical evidence, but rather repeats the general assessment of the transfer that appears throughout the literature. The restrictions on water transfers in California that he notes are part of Section 386 of the California Code. They state, in part, that a change in water use associated with a transfer may be approved only if the change can be accomplished without "unreasonably affect[ing] the overall economy of the area from which the water is being transferred."[27]

Ellen Hanak and Caitlin Dyckman suggest that the "notorious precedent" set by the large-scale export of groundwater from Owens Valley to Los Angeles that began in the 1920s was not lost to counties that imposed county-level ordinances restricting groundwater transfers following the 1977 drought.[28] The fear of unregulated groundwater mining and the consequences for rural communities and their local economies also led to state legislation in the early 1980s to restrict direct groundwater exports out of a watershed. Initially, some county ordinances were overruled on the basis of state preemption (in Inyo and Nevada Counties), but in 1992 Tehama County won an appellate decision upholding the authority of the county to regulate groundwater transfers.[29] Since then, nineteen counties introduced or regularized ordinances restricting groundwater exports, bringing the total as of 2003 to twenty-two counties with such ordinances in place. While Hanak and Dyckman note that such ordinances are a rational initial response to the lack of protection from groundwater mining for local groundwater users under the state's "no injury" rule, they also argue that the ordinances may "overprotect" the resource and impose unnecessary and inefficient barriers to water transfers that would not involve serious overdrafts of the resource.[30]

Robirda Lyon looks at California county controls on the transfer of surface waters. Drawing principally from the historical account provided by William Kahrl, Lyon attributes the promulgation of the 1943 Area of Origin laws to the concern generated in rural areas to the protracted conflict between the city of Los Angeles and Owens Valley.[31]

Hal Rothman describes in more colorful terms the social foundation for the fear and hostility engendered in rural communities by the idea of reallocating water rights to urban areas.

> Reallocation inspires deep-seated passion, for it suggests an administrative rewriting of customs of the past one hundred years. Like the farmers of the Owens Valley, who took out an advertisement that read, "We who are about to die salute you," as Los Angeles drained them of their water in the 1910s, the rural West of today sees in the shift of water to urban use a kind of social genocide that renders them not only irrelevant, but destitute. They argue for their viability with "culture and custom" arguments, romantic figments of their imagination that tugged on the heartstrings of many but demonstrated how archaic their philosophy was as well as how much power the rural West had lost. For generations, they won

this battle through careful manipulation of state and federal redistricting and by using the power they attained as the oligarchic beneficiaries of federal water to grease campaign coffers. By the late 1970s, they began to stumble, and as they lost power, a backlash against change emanated from the rural West. It argued for a perpetuation of obviously inefficient practices for the social good they created and, even more, for the extension of greater federal resources to rural areas.[32]

Andrew H. Sawyer, in his analysis of water takings issues, notes that the "reasonable, investment-backed expectations of farmers" in Owens Valley were dashed once they realized that the pattern of land purchase pursued by the city of Los Angeles would undermine the maintenance of irrigation ditches previously undertaken by private cooperatives. Without citing evidence of the practice, Sawyer argued that once some farmers along a ditch were bought out, the remaining farmers were forced to sell because they could no longer maintain the ditch on their own.[33] The issue then was not whether Los Angeles paid a "fair" price for the land, or even whether the price was substantially above market prices, but rather that the element of compulsion introduced by the large-scale land purchases forced some if not all landowners to take what they could get since they would be unable to pursue their initial objectives for investing in the land.

Joseph Sax addresses potential conflicts between private and public goods aspects of water and third-party effects from transfers. He defines "water in place" as "a type of wealth. . . . [t]hat . . . accrues not only to the owner of a water right, but to many other people in the place where the water is located."[34] The values to nonowners of water in place include employment, both direct and indirect, lower prices for water due to its relative abundance, and recreational uses. When water in place is transferred out of the community, the owner-sellers may benefit handsomely financially while the nonsellers who once benefited from the wealth of water in place will be made absolutely worse off since there is no compensating benefit to them from the transaction. Sax concludes that there is a "first-order conflict" between the holders of water rights who benefit from the sale of water in place and all the natural, social, and economic interests of all those who once shared in the wealth of the water and receive no benefit from the sale.

Sax argues that simply comparing the short-term tradeoffs between competing uses for the water—higher-value urban use versus lower-value agricultural use—does not capture the full economic and social cost of water transfers for the community in the area of origin. Even if the negative consequences for the value of town lots and commercial activities or for local employment in communities that formerly depended on the agricultural economy are compensated through the purchase of properties or the provision of relocation allowances for displaced workers, the compensation is incomplete. The future expectations and options available to communities that have lost their control over and access to water resources remain severely limited. Sax suggests that the associated fear of loss of local control resulted in the area-of-origin laws in California legislation and in other states and local jurisdictions.

By referencing the example of Owens Valley, Peter Nichols and Douglas Kenny emphasize similar concerns in discussing local resistance to water transfers from the western slope of the Rockies in Colorado. They point out that trans-basin diversions, from the relatively lightly populated and agricultural western slope to the metropolitan eastern slope, have been a battlefield for many years, "largely due to real and potential negative impacts borne by the basins of origin."[35]

> Trans-basin diversions remain highly controversial and unpopular outside the urban Front Range. Part of the reason that modern targets of trans-basin diversions view proposals with such hostility is the legacy of bitterness and distrust arising from past trans-basin diversions that remind many Western Coloradoans of California's Owens Valley. Given the growing imbalance in political and economic power, West Slope interests increasingly see themselves as being at a strategic disadvantage in these intrastate water wars.[36]

The authors note that in the case of Colorado the parties opposing trans-basin diversions have not sought to engage the receiving regions in negotiating more favorable terms and protections for water diversions, but have instead sought to derail these water projects in the state's water court.

These and other articles in law journals confirm the overall negative perception of the Owens Valley–to–Los Angeles water transfer experience. The image of the beleaguered rural community dominated by distant—and thirsty—urban interests has been almost uniformly adopted by the modern press and academics alike, if not for its truth value, then at least and perhaps most importantly, for its metaphorical power. The legend of Owens Valley is held out by communities of origin as a talisman against urban predation on rural water supplies. Proponents of water transfer also make use of the image, by defining their project in terms that make it clear that the result will *not* be another Owens Valley. The example set by Owens Valley has become the guide of what to avoid in acquiring water from a distant source. Greg James, water chief of Inyo County, the county where Owens Valley lies, characterized Owens Valley in these terms: "It's a classic example of what can go wrong with a water transfer. . . . It's the classic example. You can ruin an area."[37]

THE "OWENS VALLEY SYNDROME"
IN THE POPULAR PRESS

The story of how Owens Valley in the Eastern Sierras lost its water in the first decades of the 1900s to the growing population of urban Los Angeles over 250 miles away remains a vital part of both the history of the settlement of the West and the contemporary struggle over access to water today. While the story resonates most forcefully in the dry regions of the western states, the fate of Owens Valley has been invoked as a warning to rural communities throughout the country faced with burgeoning urban populations whose expanding demand for water exceeds available supplies.

As in the previous section, this review of the treatment of the water transfer between Owens Valley and the City of Los Angeles is based on news reports from popular newspapers and selected periodicals carried in the Westlaw "ALL-NEWS" database from 1985 to 2004.

Theft of the Valley's Water

Los Angeles's purchase of land and water rights from landowners in Owens Valley in the early decades of the twentieth century is most commonly viewed as a "theft" of water from unwitting rural people by an arrogant city.[38] Somewhat surprisingly, the principal proponents of this perspective appear to be the Los Angeles newspapers.[39] Indeed, the rags-to-riches story of William Mulholland, the Irish immigrant who rose from cleaning ditches to building a canal to transport Owens Valley's water from the eastern slope of the Sierras to the growing city 250 miles away, transforming the San Fernando Valley into an agricultural paradise in the process, has become part of the cherished urban folklore of Los Angeles. As a *Los Angeles Daily News* story commented, "The story of the Owens Valley has taken on mythic proportions in the lore of the West, symbolizing the arrogance and disdain of the powerful cities toward the rural areas."[40] The theft of the water is noted in a 1986 editorial in the *Los Angeles Times*: "The annals of Western water development teem with perfidy and connivance, but the Owens Valley will always enjoy a special aura of infamy. Mention the words water and theft in one breath, and someone will follow with Los Angeles and Owens Valley in the next."[41]

A 2001 editorial in the *Los Angeles Times* once again referred to allegations of water theft: "For more than two decades, state policy has been to encourage urban water districts to buy water from farmers. But nearly a century after the city of Los Angeles stole the water of the Eastern Sierra, the words 'Owens Valley' are invoked whenever someone talks about the city going after a farmer's water."[42] Similarly, in October 2003, a *Los Angeles Times* editorial commented: "The handsome price Imperial farmers are to receive (from the DOI [Department of the Interior] compelled settlement with IID [Imperial Irrigation District]) will pay for more efficient irrigation methods as well as for the fallowing of some land, the saved water going to San Diego. The farmers are afraid the cities will just keep coming back when they need more. However, no one is going to 'steal' the water the way Los Angeles took the Owens Valley's supply in the early part of the last century."[43]

Allegations that the valley's water was stolen also commonly appear in other newspaper articles on water issues.[44] A 2003 article in the *Economist* on water pointing to the "Owens valley syndrome" shows how persistent the story remains.[45] Besides theft of water, other themes regarding Owens Valley and its relationship with Los Angeles that appear in the popular press include neocolonialism, environmental degradation, and loss of the rural way of life.

Neocolonialism and the Loss of Water Rights

A recurring theme in the news coverage of the Owens Valley water transfer is the community of origin's loss of power and its political and economic subordination to the interests of the urban receiving community. This power imbalance has been likened to a form of colonialism. Since appropriative rights to withdraw surface and groundwater were transferable with the land and riparian rights were appurtenant to it, the LADWP acquired vast tracts of ranchland as a means of securing water rights to the Owens River and its tributaries. Gradually, the department came to own over 90 percent of the valley's lands. By buying up property, Los Angeles seized political and economic power from the local community, as emphasized in this 1985 editorial in the *Los Angeles Times*:

> Down through history, cities and city-states have gone far abroad to develop colonies in order to exploit their precious natural resources—gold, spices, tea, oil and the like. For the past 70 years the City of Los Angeles has had its own colony in the form of Inyo County—300 miles to the north, along the eastern scarp of the Sierra Nevada. . . . Inyo County residents, as a result, were left with precious little control over their own destiny.[46]

The neocolonial aspect of the relationship was also the focus of a 1986 article in the *San Francisco Chronicle*:

> Seventy years ago, the city of Los Angeles took most of the valley's water, bought nearly all the land, killed off its agricultural industry, bulldozed the farms and

turned it into an economic satellite. . . . "The people here feel like they are a colony of Los Angeles," said Gregory James, manager of the Inyo County Water Department, "but it is more like a Third World country where the principal resource is exported for the benefit of some other region. . . . Water controls the life of the valley, and Los Angeles controls the water."[47]

Another variation on the theme of water transfer as colonialism was provided in the *Arizona Trend*: "In the extreme, the advent of Arizona water farming conjures up images of the nation's most notorious hydrocolonialism, the sacking of Owens Valley."[48] The press has also stressed the dominance of Los Angeles in determining how the property it owns will be used: "It's virtually impossible for anyone to do anything in Owens Valley that involves business or ranching or public projects, without at some point having to get permission from the City of Los Angeles. . . . There's no private property and there's no water."[49]

Associated with this view of colonial powerlessness is a sense that, once even small-scale water transfers take place, politically powerful urban interests will forcibly widen the transfer despite opposition from the community of origin. Once again, referring to Owens Valley, Imperial Valley farmers voiced concern that once their water began to flow to Los Angeles it could not be stopped. This fear was given voice by Imperial Irrigation District president William Condit in a 1985 *Los Angeles Times* article: "We want to see the people on the coast get the water they need. . . . But our people look at L.A. and what do they see? The Owens Valley. They're afraid. After Metropolitan's [Metropolitan Water District (MWD)] had that water 40 years, they're not going to give it up without one helluva battle. What chance do 100,000 people have against several million?"[50]

In a *Washington Post* news story later the same year, the distrust of Imperial Valley farmers for a water trade proposed by the MWD involving the financing of water-saving infrastructure in exchange for access to part of the Imperial Irrigation District's (IID) Colorado River allocation was manifest in this comment by an IID board member:

> "The Metropolitan Water District keeps saying they're not L.A., but urban water interests are all the same. It's like comparing Tamerlane with Attila the Hun. . . . The Owens Valley syndrome is stronger than I thought it was," said board member [John] Benson. "We all think we would have one hell of a time getting the water back," Benson said. "We may get the money, but we'll never get the water back. And some day it may be more valuable to have the water.
>
> The farmers' caution is rooted in their fear that when the water is gone, it's gone."[51]

Owens Valley was used as a reference in another case where farmers feared the political and economic power of urban areas. In this case, California ranchers voiced concern over plans by the city of Reno, Nevada, to begin pumping groundwater from the Honey Lake Basin aquifer that straddles the California-Nevada border. According to a *Los Angeles Times* editorial: "Fred Mallery, a Lassen County cattleman, told the *San Francisco Examiner* recently: 'We could be the next Owens Valley if we don't stand up and do something for ourselves.'"[52]

Similarly, a MWD plan to establish a water bank in the San Joaquin Valley for use during drought was also met with distrust by valley farmers: "Among those most suspicious of the plan are farmers in the district. John Waters, president of Nalbandian Farms in Arvin, compares the agreement to 'making a pact with the devil.' He recalls the time in the early part of this century when the Los Angeles Department of Water and Power bought land in the Owens Valley and diverted the water southward: 'We think it's a bad precedent. We feel it will compromise our water rights. Ultimately, they will be our competitors for agricultural water,' Waters said. 'In a nutshell, we feel that if we don't stand up for our rights, the San Joaquin Valley could become the Owens Valley.'"[53]

Explaining concern by Nevada politicians over a California plan to divert excess Colorado River water from high water flows into the Salton Sea, California State Representative George Brown acknowledged that the state's history in regard to water transfers made it difficult to assuage the fears of neighboring states: "What you're seeing here is a historic paranoia that California is going to try to grab extra water. . . . It's like the people up in Owens Valley. . . . They're all scared to hell of L.A. because we once stole water from them. Most of the (river) basin states are afraid of California because we've stolen water from them over the years."[54]

A related colonialism concern is the lack of land in Owens Valley for economic development. Since the LADWP owns most of the property as a source of water for the city, other uses are limited. Environmentalists generally applaud the lack of development in the region that maintains its rural character, but others in the valley lobby for more development opportunities.

Environmental Degradation

Chapters Six and Seven examine conflicts over water diversion by Los Angeles in the Mono Basin and the drying of Owens Lake in Owens Valley. Here, some of the issues as presented in the popular press are summarized because they reflect the continuing contemporary relevance and legacy of Owens Valley for water market development.

Concern about air pollution arising from the dry Owens Lakebed attracted national attention. The issue involved a residue of highly volatile alkaline dust that could be picked up by spring winds that scraped across the lakebed: "Pollution experts estimate that the dust could affect at least 40,000 people in the Owens Valley," according to a 1989 *Dallas Morning News* story:

> Readings in the Owens Lake area have exceeded 1,800 micrograms of dust per cubic meter—more than 10 times the safety limit set by the U.S. Environmental Protection Agency and three times the "significant harm level." . . . If similar dust levels were found inside a factory . . . federal regulations would require that workers wear respirators. . . . Increasingly, such environmental repercussions are being considered as Los Angeles continues its quest for water, already legendary in Western states.[55]

Court-ordered measures to reduce dust storms off of Owens Lake sparked the following summation of the environmental consequences of the Owens Valley water transfer:

> The water wars of the Eastern Sierra are storied. . . . There were connivings, hard feelings, threats of violence, but it came to pass. And in due course the whole Owens Valley fell under thrall to the thirsty, hard-handed metropolis to the south. Creeks dried up, rivers dried up, even Owens Lake dried up. A vast agricultural region withered and browned, a natural ecosystem died—and an ecological monster was created. That was the dry bed of Owens Lake.[56]

This is hyperbole, however. It is not based on assessments of systematic scientific evidence of the harm caused by Owens Lake dust pollution, of which there are very few. Rather, it reflects the mythical proportions that the alleged social consequences of Los Angeles's so-called theft of the region's water have assumed for those who seek to redress what they view as a historical injustice.

A Seattle journalist, Mark Trahant, was reminded of the fate of Owens Valley when the Cedar River, Seattle's primary source of water, was placed on the top-ten list of the nation's most endangered rivers by the conservation group American Rivers. Trahant characterized the dry bed of Owens Lake as "where 20th-century technology stole a river system and redesigned it to flow through city faucets." Concerned by the city of Seattle's plan to double its withdrawal from the river and what he saw as the city's unending thirst for a precious resource, Trahant warned, "We have either learned from the past century or, if not, we're doomed. We can balance the natural world with urban needs or declare the Cedar River an urban sacrifice area just like the Owens Valley."[57]

In *Sierra*, a publication of the Sierra Club, Las Vegas and Los Angeles were compared in their unending search for water. Referring to the explosive growth of southern Nevada and the attendant search for new sources of water, an article equated the profligacy of water use in Las Vegas with "a type of environmental terrorism":

> What Las Vegas cannot buy from Arizona farmers, it seems determined to divert from the Virgin River (a tributary of the Colorado) or steal from the ranchers in Nye and Lincoln counties. Over the next decade, it may desiccate central Nevada and southwestern Utah as thoroughly as Los Angeles did the once-lush Owens Valley on the eastern flank of the Sierra, when it stole its water 80 years ago (an act of environmental piracy immortalized in the film *Chinatown*).[58]

Loss of a Rural Way of Life

The press also views the Owens Valley experience as an example of the destruction of a rural way of life, the loss of cultural values as local farmers and ranchers were driven from the land and obliged to take up new livelihoods in urban areas. Recall that historian William Kahrl stressed the agricultural promise of Owens Valley that was lost when Los Angeles took the water.[59] Without any serious examination of the real agricultural potential of the region, the view of a "paradise lost" remains a powerful one in contemporary accounts of the Owens Valley transfer: "Rural towns remain suspicious of any water sales that could turn their agricultural areas into a modern-day version of California's Owens Valley, still shorthand in Western water circles for economic rape and pillage."[60]

Indeed, restrictions on water sales to protect and preserve rural communities against the growing water demands from urban areas are supported by rural politicians. Even though farmers pay just a fraction of the amount per acre-foot that a municipal authority is willing to pay for water, Republican Idaho senator James McClure called for regulation of water markets "so rich outside interests cannot overwhelm the social and economic structure of a community.... There is no question Idaho could look like Owens Valley if we allowed California to buy our water."[61]

The fear, repeated in the press, is that if water is transferred to urban areas, the local rural agricultural economy would be ruined. A 2001 article in the *San Jose Mercury News* claimed: "But water transfers, even when voluntary, arouse great fears from rural residents. They often cite Los Angeles' hardball push 90 years ago to buy up water rights in the Owens Valley near Bishop, a tactic that destroyed the agricultural economy there."[62] And of course, after the water was taken from the valley, it did not go to just any community, but to Los Angeles, the city that epitomizes urban sprawl in the minds of many environmentalists and journalists: "Just south of where Nevada sticks its elbow into California's belly is the Owens Valley. From here, decades ago, Los Angeles stole the water that supported southern California's thoughtless population expansion. Today, there's often more water flowing through the canals that feed L.A.'s swimming pools than there is in the Owens River."[63]

Commenting on a plan before the Colorado state water court to pipe water from an aquifer in the San Luis Valley on the western side of the Rockies to urban areas on the eastern side, San Luis Valley potato farmer Jim Tonso Jr. stated, "The impact would be disastrous. . . . It would lower the water table and we couldn't economically farm anymore. It would be the same as the Owens Valley."[64]

SOME EQUAL TIME: POTENTIAL SOCIAL GAINS
FROM RURAL-TO-URBAN WATER TRANSFERS

Although the law review and popular press accounts of the Owens Valley transfer summarized here stress the negative aspects of the exchange and the importance of protecting and subsidizing rural lifestyles and community values through restrictions on water trades to cities, it is important to recall the social values of urban areas and the need to provide them with sufficient water.[65]

The most rapid population growth in the United States is in the urban areas of the semi-arid West. This growth is fueled by shifts from agriculture and extractive industries to service and technology industries. Most western cities, including Los Angeles, Las Vegas, San Diego, Phoenix, Denver, and Tucson, do not have sufficient local water sources to supply this growth in urban demand. Accordingly, meeting their growing demands provides the impetus for most efforts to acquire water from agriculture, where most western water currently is used. Given how much water is applied to agriculture, moving relatively small amounts of it in any one area may be sufficient to address major portions of urban demand. The almost complete absorption of a region's water, as took place in Owens Valley, is unlikely to occur in contemporary water transfer efforts. Accordingly, the economic and environmental third-party impacts are likely to be much more limited. In any case, there may be considerable social gains from providing more relatively low-cost fresh water to meet growing urban demand.

In general, both historically and today, cities are the location of much of the economic development and innovation that occurs in the country.[66] Cities typically offer the highest real wages, and are the location of most opportunities for productivity growth, employment, and upward mobility; hence their attractiveness to the poor.[67] Accordingly, most migration patterns are from rural to urban areas.[68] The economic advantages of cities all derive from increasing returns or agglomeration externalities. Of these, there are three broad categories: technology, public goods, and consumer amenities. In the case of technology, cities facilitate the flow of knowledge and allow for markets to emerge around new innovations. This explains why inventors, innovators, and entrepreneurs have concentrated in urban locations to develop and market their ideas.[69] In the case of public goods, such as transportation and education, there are often high fixed costs that can be better spread in urban areas. This situation promotes investment in public goods in cities that might not be cost-effective in rural areas.[70] Finally, in the case of consumer amenities and market size, in urban areas the search costs for new products and processes, as well as for jobs, are lower because of the clustering of retailers, producers, and employers.[71] For these reasons, consumers looking for goods and services, and workers looking for employment, flock to cities.

As cities have expanded, they have spread spatially with different density gradients. While this is referred to pejoratively as urban sprawl, it is characteristic of all city growth both today and historically.[72] If transportation costs were to rise in comparison with past patterns, future urban development would be denser. As noted above, surely one reason why Owens Valley carries such cynical notoriety about rural-to-urban water transfers is that its water was the basis for the growth of the city most associated with sprawl—Los Angeles. Yet urban demands for water are growing throughout the West, and some reallocation of water will occur. The objective is to facilitate water transfers that will be as smooth and low-cost as possible, both in direct costs and in third-party effects.

CONCLUSION

The Owens Valley–to–Los Angeles water transfer clearly plays an important role in molding how the reallocation of water from rural to urban areas is viewed through both the academic and popular presses. A recent Associated Press story on Colorado farmers selling water to the Denver suburb of Aurora ran in eleven cities across the nation, including Washington, D.C.; St. Paul and Duluth, Minnesota; Los Angeles, San Diego, and Oakland, California; Aberdeen, South Dakota; Charleston, West Virginia; Ft. Worth and Houston, Texas; and Tulsa, Oklahoma. The article reports on the development of water markets in the West, with reference to the Imperial Valley, Owens Valley, and Albuquerque and Texas water markets, along with the lead story on water transfers from Colorado's Arkansas Valley to Aurora.

The news story presents what it terms "a cautionary example of the West's burgeoning water market" and relates the consequences for the farmers and communities of the Arkansas River Valley of the sale of their water to Aurora for a one-time payment of $18 million. As an example of farm towns that have "gone under" when they sold their water to cities, the article relates how Owens Valley "became a dust bowl when Los Angeles *quietly* acquired its water and flushed it down an aqueduct to the city 90 years ago."[73]

As Chapter One details, there are important economic and political pressures for the movement of water from low-value agricultural uses to higher-value urban and environmental uses. Unlike Owens Valley, most contemporary water movements will involve a comparatively smaller share of a region's water. And the economic and social benefits from this reallocation are very large, as indicated by the price differences for water at the margin in urban and environmental uses, relative to agricultural irrigation. Yet even these exchanges are viewed with suspicion, colored by what people believe to be the Owens Valley experience.

As with most popular mythologies, some aspects of the Owens Valley story contain elements of fact. Other aspects are clearly myths that have emerged as the story of Owens Valley has grown, used by various parties for their own purposes. The common notion that Owens Valley water was stolen only has meaning for the distribution of the gains of trade. The analysis in Chapter Five shows that although farmers in Owens Valley were made better off by selling their water and land than if they had remained in agriculture, the majority of the value of the water was captured by Los Angeles. If an unequal division of the benefits of the exchange of water and land is termed "theft," then theft occurred. But if "theft" is meant to describe the plight of unwitting and unwilling farmers who lost their water and land to Los Angeles, as is commonly portrayed, then no theft occurred.

Owens Valley was a marginal farming area with comparatively low agricultural returns in comparison with other Great Basin regions and the rest of California. Most farmers appear to have been delighted with the opportunity to sell their properties and to sell to a rich city like Los Angeles. They were not "unwitting," and Los Angeles did not "quietly" or secretly buy up most of the properties in the valley. Rather, Owens Valley farmers shrewdly attempted to organize as a single bargaining group to increase their negotiating power with the LADWP in the early 1920s when the city began to buy up the most valuable farmland. When this effort failed, smaller, more cohesive groups founded separate sellers' pools to negotiate with the city. And those farmers who were members were able to sell their land for a higher per-acre price. But they were not able to raise their prices enough to capture more of the value of the water transferred with their land. Most of the history of Owens Valley as described in the first part of this book, then, is one of negotiations between two groups, farmers and town lot owners in Owens Valley and representatives of the LADWP, each trying to do the best it could do in land and water sale negotiations.

There was no devastation of the Owens Valley economy, nor was the region turned into a desert or "dust bowl," although the export of its water did have negative environmental effects. The nature of the economy gradually shifted, but as in other Great Basin regions, it would have shifted from small farms to larger ranches. Had Owens Valley not had a pristine mountain environment (much of it preserved by Los Angeles) and close proximity to Los Angeles, then its small towns would have declined in population more than they did. By buying up virtually all of the properties in Owens Valley, Los Angeles internalized many of the physical externalities associated with the diversion of Owens River water to the aqueduct and with groundwater pumping.

3

THE HISTORY OF THE OWENS VALLEY–TO–LOS ANGELES WATER AND LAND EXCHANGE

The City of Los Angeles is interested in buying water. It is necessary to buy the land to get the water.

Cope Rand Means Co., Engineers, 1923

This chapter provides an overview of the history of the Owens Valley–to–Los Angeles water and land exchange. The material presented here provides the background for subsequent analysis of the negotiations between landowners and representatives of the Los Angeles Department of Water and Power (LADWP). It draws on the very thorough historical accounts provided by Abraham Hoffman, William Kahrl, Vincent Ostrom, Robert Sauder, Remi Nadeau, and others. Many of the historical details involving people central to the Owens Valley story, such as William Mulholland, Joseph B. Lippincott, Fred Eaton, W. B. Mathews, and Wilfred W. Watterson, are omitted here in an attempt to identify major patterns and incidents in the bargaining history between valley property owners and representatives of the LADWP.[1] The agency started as the Board of Water Commissioners and in 1910 became the Board of Public Service Commissioners, Bureau of Water Works and Supply. In 1915 the Bureau of Power and Light was added, and by 1931 it was called the Board of Water and Power Commissioners. The Board of Water and Power Commissioners establishes policy for the LADWP, and the two bodies are referred to interchangeably throughout this volume.

Analysis of the bargaining between the parties identifies why the negotiations were so contentious, leaving such a negative legacy, and shows how the parties fared in the process. Table 3.1 provides a chronological list of events as they emerged between 1900 and 1935 in the relationship between the LADWP and the farmers and townspeople of Owens Valley. For an otherwise remote (although beautiful) valley in the eastern Sierra Nevada Mountains, this is a remarkable history in its importance for understanding the economics and politics of western water today.

TABLE 3.1
Chronology of Los Angeles–Owens Valley land and water negotiations

Year	Event
1900	— The City of Los Angeles's population reaches 102,497, up 103 percent from 50,395 in 1890; the county population increases to 170,298, up from 101,454 in 1890.
	— Concern grows about the water supply as a threat to the city's growth.
1903–09	— Former mayor Fred Eaton acquires over 50,000 acres in the southern, nonirrigated part of Owens Valley, as well as a northern reservoir site. He purchases land secretly and plans for joint water and power development with Los Angeles.
	— The LADWP quietly buys 22,670 acres of this land and the water rights from Eaton, making the city the residual claimant for water not used in irrigation.
	— On July 29, 1905, the Los Angeles Times announces the acquisition of Eaton's properties and the future arrival of Owens Valley water.
	— On September 7, 1905, a bond issue for $1.5 million is approved for land purchases and an aqueduct survey.
	— The campaign for approval of the bond issue is contentious, with the motives of aqueduct supporters challenged as land speculation. Private water and power companies are also opposed, but the Chamber of Commerce supports it.
	— The Federal Reclamation Service considers, surveys, and then abandons an irrigation project in Owens Valley that was supported by local residents. Residents charge conflict of interest among Reclamation personnel. The city seeks some of the same lands and water.
	— Los Angeles obtains federal right-of-way for an aqueduct with a requirement that the project must be public. President Roosevelt supports the transfer of water to Los Angeles for "greatest good" uses. The episode generates resentment in the valley toward the city.
1907–12	— In 1907, voters approve a $23 million bond issue for construction of the Los Angeles Aqueduct.
	— Los Angeles purchases more lands in southern Owens Valley, mostly from the federal government, increasing its ownership to 82,000 acres. There is little controversy.
	— The Los Angeles city population rises to 319,187 in 1910, an increase of 211 percent from 1900.
1913–20	— The aqueduct completed in 1913 is 250 miles long and is a major public works project.
	— In 1913 another water bond passes for $1.5 million.
	— The L.A. Aqueduct delivers four times the domestic demand, with most water initially directed to San Fernando Valley land. Annexation of San Fernando Valley follows. Land values rise dramatically, from $20 to $2,000/acre, fueling charges of land speculation.
	— There is limited negative impact on agriculture in Owens Valley.
	— Aqueduct flow is vulnerable to water diversion by ditches above the intake. Subsequent negotiations between the LADWP and farmers to maintain irrigated land in Owens Valley, to construct a water storage site, and to guarantee the water flow in the aqueduct fail.
	— The Owens Valley Defense Association is formed to represent irrigated areas. A competing group, Associated Ditches, supports the city's plan to guarantee flow and irrigation.
	— The Los Angeles city population rises to 576,673 in 1920, up 81 percent from 1910.
1921–22	— A $5 million water bond for Owens Valley lands passes in 1922.
	— San Bernardino v. Riverside, 186 Calif. 7, allows Owens Valley landowners to block pumping of groundwater.
1923–24	— A serious drought begins; aqueduct capacity is 400 cubic feet/second, but actual flow is 260 cubic feet/second.
	— Los Angeles pumps groundwater in its southern properties to fill the aqueduct. Injunctions are filed, and the city purchases adjacent properties to maintain pumping.

TABLE 3.1
(*continued*)

Year	Event
	— The LADWP begins buying land in northern Owens Valley. There are price disputes over some properties. The aqueduct is dynamited. State and national press coverage of "California's little Civil War" is critical of the LADWP.
	— Owens Valley and Big Pine Reparations Associations form to demand compensation for deterioration in the agricultural economy. They appeal to state and national press.
	— The Owens Valley Irrigation District (OVID) is organized to consolidate the water rights of major ditch companies and to negotiate as a unit with the city. The city counters with purchases of critical farms and the district fails.
	— Three sellers' pools form.
	— Water bonds for $8 million are passed in 1924.
1925	— The LADWP drops offer to retain 30,000 acres of irrigated land and creates a Board of Appraisers to value properties and to set offer prices.
	— The Packard Report recommends that the city buy up remaining properties to guarantee water supply and that an outside group set prices.
	— A Special Owens Valley Purchasing Committee is created by the LADWP to negotiate with the Owens Valley Land Owners Committee.
	— Damage claims of $2,813,355 are presented by the Owens Valley and Big Pine Reparations Associations.
	— State legislation makes cities liable for losses due to removal of water from the watershed. The LADWP contests its constitutionality.
	— Aqueduct and city wells are dynamited, and periodic diversion of city water continues.
	— Water bonds for $8 million pass.
1925–29	— Los Angeles actively purchases more farms, but negotiations are contentious.
	— Members of the Owens Valley Ladies Committee testify before the LADWP that the agency is not paying "fair" value for their properties. They argue that the Appraisal Board is biased.
	— The Owens Valley Owners Protective Association forms and adopts aggressive publicity campaign against the LADWP. Ads are placed in leading state papers to stress the plight of small farmers and towns against Los Angeles. "We Who Are About to Die" appears in the *Sacramento Union*. The critical series runs March 28–April 3, 1927.
	— Investigations by the LADWP, the mayor, the governor, and the California Assembly seek to resolve controversy over land purchases.
	— The LADWP halts purchases of property in Owens Valley between 1927 and 1929.
	— Water bonds for $10 million pass in 1926.
	— The LADWP begins negotiations to purchase remaining farm and town properties. A Special Owens Valley Committee (three board members) negotiates with the Committee of Ten (two representatives from each of five towns). Contentious negotiations follow.
	— Some Los Angeles taxpayers criticize high payments for Owens Valley properties.
1930	— The Los Angeles city population reaches 1,238,048, a gain of 115 percent from 1920.
	— A bond issue for $38.8 million is passed to acquire town properties and private power company lands, and to extend the system to the Mono Basin.
	— Negotiations for town properties continue. Sellers' pools form (Main Street Pool).
1931–35	— The LADWP completes its purchase of town properties.
	— A state senator from Bishop represents the Main Street Pool and calls for Senate investigation of the LADWP purchasing practices. A critical report is released.
	— More injunctions are threatened against pumping on city lands, and the aqueduct is dynamited.
	— By 1935, Los Angeles owns 96 percent of farm properties and 85 percent of town properties.

THE AQUEDUCT AND EARLY LAND PURCHASES

As outlined in Table 3.1, the Owens Valley story begins with Los Angeles's search for more water. Between 1890 and 1900 the population of the city of Los Angeles more than doubled, to 102,497 people, and the county gained even more new inhabitants. Given the city's climate, links to the East via the intercontinental railroads, and position as a major West Coast port, prospects for continued growth seemed promising, except for the absence of sufficient water.[2] The city was in a semi-arid region where annual precipitation not only was extremely variable, but averaged just 14.62 inches, whereas Chicago, for example, had mean rainfall of 34.12 inches.[3] Los Angeles relied on the meager Los Angeles River watershed, rather than rainfall, for its water supply. But by the turn of the century, there was growing concern among city boosters that more water had to be found if the city were to achieve prominence as the leading city on the West Coast.[4] And there was water, 250 miles northeast in Owens Valley on the eastern slopes of the Sierras. Between the Owens River's flow and ground sources in the valley, there were some 37 million acre-feet (a.f.) of water available, about the same as that held in Lake Mead today.[5]

Owens Valley: A Water, Not an Agricultural, Wonderland

Owens Valley is an elongated trough, lying between the Sierra Nevada and Inyo Mountains, about 120 miles long and varying in width from two to six miles. Both bordering mountain ranges are high, at 10,000 feet or more, and indeed, the Sierras in this area are the highest in the lower forty-eight states of the United States, with Mount Whitney looming above the valley. The Sierras capture significant precipitation and there is heavy discharge into Owens Valley. The valley is bisected by the Owens River, which, until it was diverted into the Los Angeles Aqueduct, dumped into the alkaline Owens Lake. The valley is filled with porous material that absorbs most of the annual runoff from the mountains, except in very wet years. There is no outlet for Owens Valley water, and groundwater levels are close to the surface, creating artesian conditions.[6]

In 1920, before major land purchases by Los Angeles, there were 7,031 people in the area on farms and in five small towns—Bishop, Big Pine, Laws, Independence, and Lone Pine.[7] There were 140,000 acres of farmland in the valley, of which 39,904 acres were improved, with perhaps 30,000 in irrigation.[8] Livestock was a principal agricultural product. The elevation of the valley (ranging from 3,600 to 4,300 feet), short growing season (150 days), alkaline soil, and limited access to markets constrained its agricultural potential, an issue examined in

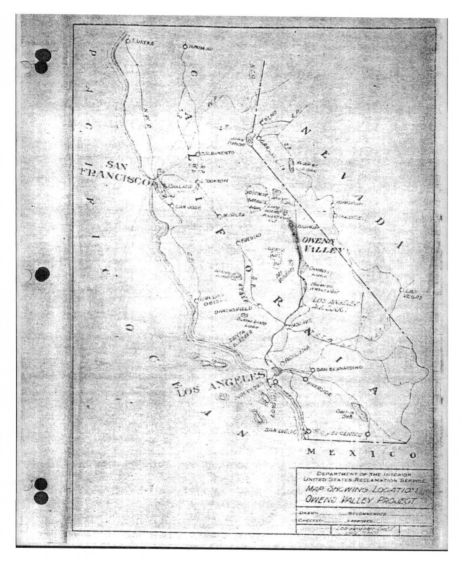

Figure 3.1 Owens Valley.

source: Reprinted from Conkling (1921).

more detail later. Figure 3.1 shows the location of Owens Valley northeast of Los Angeles.

There were two types of water rights in Owens Valley—riparian and appropriative. Riparian rights were more limited, and they granted each landowner along the Owens River or adjacent streams the right to the reasonable use of water appurtenant to their land. The more dominant appropriative rights granted land-owners, generally those whose properties were not along a stream, an amount of water that could be transported via a ditch or aqueduct and placed in beneficial use. Groundwater also was available, and surface property owners could pump groundwater also for reasonable use.[9]

The distribution of water in Owens Valley was not uniform. There were vast amounts of water in certain parts of the valley, especially in the northern area around Bishop, while other parts were comparatively dry. A constraint on agri-culture in the northern, wetter section of the valley was an inadequate amount of farmable land and a high water table. The valley floor in that region was bro-ken up by small hills, ridges, and rock formations. To the south, much of the land was higher and drier. With limited arable land in the north, not all of the water in well-endowed areas could be absorbed in the available cultivatable acre-age and contribute to agricultural production. As a result, much of it was used wastefully in flood irrigation. Indeed, geologists Charles H. Lee and S. B. Rob-inson, who visited the valley in 1910 to assess its agricultural production and to calculate the water supply, commented on the profligate use of water by Owens Valley farmers that left some lands waterlogged.[10]

Similarly, Harold Conkling's 1921 report on Owens Valley for the Reclama-tion Service indicated that around Bishop there were 22,000 acres in cultivation, 54,000 acres potentially irrigable, and 41,200 acres deemed as "seeped," too wet to cultivate because they had heavy deposits of salts. To the south, around Inde-pendence, 5,700 acres were cultivated, 58,600 potentially irrigable, and 28,500 acres "seeped." [11] Another engineering study conducted in 1923 also indicated heavy use of water by farmers of between 4.7 and 8 a.f. per acre of land.[12] These figures generally exceed current irrigation levels in California, which range from 3 to 6 a.f. per acre.[13]

In Owens Valley in the early twentieth century, all of the surface water rights had been claimed by farmers individually or as part of ditch companies. There was no unappropriated water. As a result, those who sought to acquire water for Los Angeles had to buy either existing water rights separately or both the land and the water on each farm. Today, farmers can sell or lease some of their water rights and remain in agriculture. In Owens Valley at the time, even in the areas with the most water, this was not feasible because the farms were so small that each one provided only a tiny fraction of what was needed for Los Angeles. The limited amount of water associated with any one farm did not justify purchas-ing only a portion of the water rights. And because the farms were economically marginal, most farmers sought to sell their land and water, rather than to release any excess water for sale. Sales of excess water would have involved additional

Figure 3.2   The Five Great Basin Counties and Los Angeles County.

measurement and enforcement problems that could be avoided if Los Angeles secured the entire farm. And this is what the LADWP did.

Census data indicate just how marginal the region was for agriculture and why farmers might have been very eager to sell to an interested buyer. Comparing Inyo County (Owens Valley) farms with a baseline of farms in similar Great Basin counties—Lassen in California and Churchill, Douglas, and Lyon in Nevada—for 1920 reveals that Inyo farms tended to be smaller on average (269 acres versus an average of 713 acres for the other four counties) and the annual value of production per farm lower ($4,759 versus $10,069). Figures 3.2 and 3.3 show the location of the five Great Basin counties in relation to Los Angeles and the overall Great Basin.[14]

As we will see in the next chapter, many of the statements made by farmers during their negotiations with the city can best be interpreted as part of a

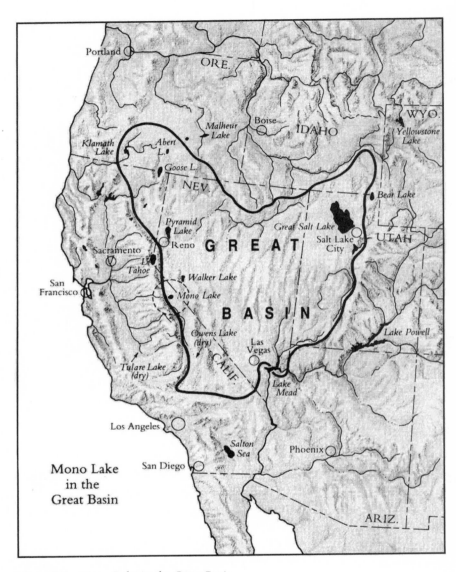

Figure 3.3 Mono Lake in the Great Basin.

bargaining strategy to hold out for a higher price. Given the small size of the farms, along with the climate, soil, and isolation of the region, serious consolidation of farms would have to have taken place in order for some to remain economically viable over the long term. This pattern of consolidation occurred in other similar Great Basin counties where no purchases were made by Los Angeles. This characterization of Owens Valley agriculture contrasts with the usual accounts that portray farmers as desiring to maintain their farms and way of life against the aggressive advances of Los Angeles.

Los Angeles Comes to the Valley, 1905–23

Between 1903 and 1905 a former mayor of Los Angeles, Fred Eaton, acquired over 50,000 acres at the southern end and northern tip of Owens Valley. Most of this land was part of the Rickey Land and Cattle Company, which included an important reservoir site in Long Valley in the far north of Owens Valley. This site also was of interest to the Federal Reclamation Service, which was considering a drainage and irrigation project in Owens Valley at the same time. Later, in the early 1920s after the Reclamation project was no longer under consideration, there were tense negotiations with Eaton over a proposal by Los Angeles to construct a high dam for a large storage reservoir in Long Valley. These negotiations ultimately failed. Before 1920, however, Eaton's southern Owens Valley lands were more critical than those in the north for providing sufficient water rights for Los Angeles to justify construction of an aqueduct between the city and Owens Valley. When the Board of Water and Power Commissioners purchased about half of the Rickey ranch from Eaton in 1905, it secured not only the land (which was incidental), but "[a]lso water and water rights and all interests, claims and appropriations in or to water rights and all interests in water ditches or canals that are in anywise appertaining to any of the lands hereinbefore in this Indenture mentioned."[15] The board also claimed groundwater lying under these properties.

Eaton acquired other smaller properties besides the Rickey ranch, and at least for some of these purchases Eaton successfully hid the true aim of his acquisitions to secure water rights in an effort to avoid a runup in prices.[16] The lands were acquired ostensibly for ranching, and when plans for an aqueduct were announced in 1905, those who had sold cried foul.[17] The Board of Water Commissioners also concealed their negotiations with Eaton: "The proceedings of the Board in these transactions were conducted with the utmost secrecy, in order to prevent speculators from anticipating the city in securing the property desired."[18] From the point of view of their customers and taxpayers in Los Angeles, this action by the board was prudent.

The General Land Office Registrar in Independence, Stafford Austin, how-ever, charged that Eaton had deceived sellers by misrepresenting his intentions when buying the properties.[19] His acquisitions certainly are the basis for the commonly asserted claim that Owens Valley lands were purchased "secretly" or "quietly." However one might judge Eaton's efforts at secrecy to keep prices low, these purchases were made very early (before 1905) and involved a very small portion of the total land subsequently acquired by Los Angeles in Owens Valley. The farm sales that took place between 1905 and 1935, which involved the vast majority of land and water rights in Owens Valley, were made with all parties fully aware of what was at stake. For these properties, the issue, as we will see, was one of price. Nevertheless, the "secret" early buying of some Owens Valley lands, along with the cancellation of a proposed Reclamation Service irrigation and drainage project, are important ingredients in historical accounts that argue that Los Angeles "stole" Owens Valley water in order to enrich San Fernando Valley speculators.

In May 1905 the board purchased 22,670 acres of land with water rights and sixteen miles of Owens River frontage in southern Owens Valley around Inde-pendence from Eaton for $450,000, or $22.50 per acre:

> WHEREAS, it has become manifest to the Board that the present water supply of the city is now scarcely adequate to the needs of its inhabitants, and, if the present rate of increase of population shall continue, will soon be insufficient to meet the demand of the City for water for necessary municipal and domestic purposes; and
>
> WHEREAS, the members of the Board feel that immediate steps should be taken to secure such additional water rights as will insure an adequate supply of water for the future needs of the city, and the Board has, with this end in view, caused to made through investigations as to the various sources of supply which are available and would afford the volume of water required by the city; and . . . NOW THEREFORE, BE IT RESOLVED, that the offer to Mr. Eaton be accepted.[20]

In 1906, the board acquired another 11,850 acres, mostly riparian to the Ow-ens River, from the William Penn Colonial Association for $121,223.[21] Through this and subsequent early acquisitions, the board became a claimant to excess Owens River water that had not been diverted by irrigation ditches in the north-ern part of the valley. Los Angeles maintained its position as a residual water claimant until the drought of 1923. As such it did not interfere with irrigated agriculture in most of the valley.[22]

As part of his property schemes, Eaton had planned for the joint development of water and power with Los Angeles. This turned out to be not possible. As occurred elsewhere in the United States at this time, Los Angeles municipalized water and power utilities under its jurisdiction, and a congressional requirement for granting the aqueduct right-of-way across federal lands was municipal ownership of the water.[23] Hence, Eaton and private power companies that owned land in Owens Valley had to sell their properties, and they subsequently became embroiled in negotiations with the board, adding to the complexity of land transactions in the valley.

Once the city's purchases of land and water rights in the southern Owens Valley were announced in the *Los Angeles Times* in 1905, successful but contentious votes on two bond issues for $24.5 million to buy land and to construct an aqueduct were held in 1905 and 1907.[24] The issues were controversial because of suspicions that land speculation in the San Fernando Valley was the primary motivation for Owens Valley water, not impending shortages. The inflation in San Fernando Valley land prices was accompanied by allegations of insider land purchases as dramatized in the movie *Chinatown*.[25] Private water and power companies in Los Angeles also disputed their buyout by the city, and there were charges of corruption related to the aqueduct's construction. A Special Investigating Committee was formed and issued its report in 1912, but there does not seem to have been much behind the charges.[26] The allegations, however, may have increased tax- and ratepayer scrutiny of the board's purchases in Owens Valley and made the agency a tougher bargaining party.

Indeed, the pending arrival of Owens Valley water brought dramatic property value gains in Los Angeles.[27] Although Los Angeles subsequently grew more rapidly than predicted, requiring more water for urban demand, much of the initial water went to irrigate lands in the San Fernando Valley.[28] When the aqueduct began flowing in 1913 it supplied four to five times the domestic urban demand, but under the appropriative water rights doctrine the water had to be in beneficial use to retain ownership.[29] Accordingly, water was made available for farming in the San Fernando Valley, and irrigated acreage in Los Angeles County expanded by over 124,000 acres. The board provided Owens Valley water only to areas that agreed to be annexed by the city, and this provision led to the dramatic increase in the size of Los Angeles by over 325 square miles.[30] Contemporary newspapers also reported that land values rose from a few dollars an acre to $500 and more per acre by 1913.[31] To illustrate the effect of the arrival of Owens Valley water, taking $300 as the average increase in land value in the nonurban areas of Los Angeles County alone, the gain was over $37 million in 1910 for the newly added farmlands and possibly an additional $113.7 million for increases in value for existing agricultural lands due to a more certain water source and greater urban land expansion.[32]

Another source of controversy was a canceled Reclamation Service development for Owens Valley. At the same time that Los Angeles was planning to acquire Owens Valley water, its residents were seeking a reclamation project for drainage of waterlogged areas and increased irrigation in dry regions. Ultimately, these were competing uses for Owens Valley water, and Los Angeles officials moved aggressively to secure access to federal lands for right-of-way and for reservoir storage sites. The agency investigated prospects in the valley between 1903 and 1905, but suspended activities in 1905 at the recommendation of J. B. Lippincott, who was both the supervising Reclamation Service engineer for California and a consulting engineer for the Los Angeles Board of Water Commissioners.[33]

Also at the recommendation of Lippincott and others, on June 25, 1906, President Theodore Roosevelt endorsed a right-of-way across public lands for an aqueduct, the withdrawal of adjacent public lands from private entry and claiming, which would have required that Los Angeles buy those properties, and the use of Owens Valley water in Los Angeles for both domestic and irrigation purposes, with the president claiming: "Yet it is a hundred or a thousand fold more important to the state and more valuable to the people as a whole if used by the city than if used by the people of Owens Valley."[34] Political maneuvering by politicians and city officials to secure cancellation of the reclamation project and to obtain the right-of-way, as well as Lippincott's apparent conflict of interest, fueled speculation in the valley that the political influence of Los Angeles had doomed the project.[35]

Much is made of this in the historical literature as evidence of the city's political power and lack of concern for the welfare of the valley's residents.[36] But recent research indicates that the Reclamation Service's decision was based on limited funds and more favorable sites elsewhere in the West, including Yuma in Arizona, the Klamath Basin in Oregon and California, and along the Sacramento River and elsewhere in California.[37] In any event, the loss of the reclamation project and the diversion of Owens Valley water to irrigation in the San Fernando Valley, where huge capital gains were earned, likely soured relations between Los Angeles and at least some farmers. The analysis described in the following chapters suggests that most farmers in Owens Valley, however, were delighted with the opportunity to sell their small farms to a party with the wherewithal of Los Angeles, so it is difficult to evaluate the real importance of the lost Reclamation Service project at the time. Some of the claims made by farmers subsequently may have been part of their negotiation posturing, which was both very political and public relations–oriented.[38]

Further, given the narrow confines of the valley, its soil, and its elevation, it is not clear how productive the venture might have been, and this seems to have influenced the Reclamation Service's decision. Certainly many of the small farms would have to have been consolidated in order to take advantage of any increased irrigation made possible by the drainage plan.[39] The sale price of farms may have increased with the project, based on the value of greater alfalfa production, but the opportunity to sell to Los Angeles, where water was even more valuable than alfalfa, likely was of greater attraction to the valley's residents. And battling over the sale values of land and water was the source of contention that made Owens Valley famous. Nevertheless, there is no question that the Reclamation Service issue has dominated subsequent discussions of Owens Valley by historians and how the Owens Valley–Los Angeles relationship is viewed today. Indeed, the two major contemporary historians of Owens Valley, William Kahrl and Abraham Hoffman, both devote approximately fifty pages to the episode.[40]

The Los Angeles Aqueduct

The Los Angeles Aqueduct was completed in 1913 for $24.6 million and was the largest public works project of its time, as well as an engineering marvel that attracted national and international attention.[41] The aqueduct was second in size only to the Panama Canal.[42] Construction of the aqueduct involved 240 miles of phone lines, 500 miles of new roads, 2,300 buildings, 1,000 men to dig tunnels, and 3,000 to construct canals, siphons, and other infrastructure.[43] New technologies, methods, and power were used, and it was covered by the national and international press, including *National Geographic Magazine*, *Sunset*, and *The Outlook*.[44] Once finished, the aqueduct was 223 miles long and included 52 tunnels, 12 miles of pipe, and 98 miles of conduit, with an additional 113 miles of canals.[45]

The aqueduct tapped the Owens River thirty miles north of Owens Lake at the lower end of the northern part of the valley below the diversion of most of the major irrigation canals in Owens Valley. This intake location, however, made Los Angeles vulnerable to subsequent illegal diversions of the water it purchased upstream for the aqueduct.[46] These diversions were part of negotiating strategies by farmers in ditch companies who were bargaining with the LADWP over the sale of their properties. The diversion of aqueduct water raised anxiety in the city about the security of its water supplies and thereby increased pressure on the LADWP to agree to the prices demanded by those farmers.

The maximum capacity of the aqueduct was between 475 and 480 cubic feet

TABLE 3.2

Los Angeles Aqueduct flow and share of urban demand, 1920–35

Year	Aqueduct flow (cubic feet/second)	Aqueduct flow (acre-feet/year)	Share of aqueduct flow in urban demand (percent)
1920	283	204,524	54
1921	263	190,070	57
1922	346	250,054	60
1923	264	194,406	59
1924	198	143,095	71
1925	270	195,129	73
1926	251	181,398	72
1927	367	265,231	74
1928	297	214,642	69
1929	268	193,684	67
1930	347	250,777	70
1931	342	247,163	70
1932	346	250,054	73
1933	341	246,441	72
1934	326	235,600	71
1935	357	258,004	72

SOURCE: Ostrom (1953, 22, 24). Ostrom reports aqueduct flow in cubic feet per second (cfs). He states that in the 1950s the aqueduct had a capacity of 450 cfs. Using the conversion of 450 × 1.98 to acre-feet/day × 365 days, the cfs = 325,215 a.f./year. The other cfs data are converted to acre-feet/year in the table.

per second (cfs), but in 1921 the practical capacity of the aqueduct was estimated at 400 cfs per day, or 289,080 acre-feet.[47] It delivered 260 cfs in 1918, 283 cfs in 1920, and 346 cfs in 1922, only to fall to 198 cfs during the drought of 1924 (Table 3.2; see photos of the aqueduct's construction in Figure 3.4). The water supplied by the aqueduct dwarfed previously available supplies to the city. For example, in 1920 the entire Los Angeles River basin supplied a flow of just 68 cfs.[48]

All in all, this was a very important water transaction. Until the arrival of Colorado River water in 1941, there were no other large sources of water available to Los Angeles. The new water made the growth of semi-arid Los Angeles possible. By 1930 the population of the city and county had surged to 1,238,048 and 2,208,492 people, respectively. As improbable as it might have seemed in 1905, Los Angeles hosted the Tenth Olympic Games in 1932. For a time, Los Angeles became the nation's largest agricultural county in terms of value of production. Associated power generation made possible by the gravity flow of Owens Valley water through the aqueduct made the Los Angeles Department of Water and Power the largest municipal electric utility in the country.[49]

The Pattern of Owens Valley Land Purchases by Los Angeles

Table 3.3 (see p. 46) outlines the purchases made by the LADWP for Owens Valley properties between 1905 and 1934 in real terms. A number of patterns emerge from the data. First, most of the early purchases in the southern, least populated, and more desert part of Owens Valley around Independence were completed between 1905 and 1910, and the mean purchase price was relatively low. Eaton's properties accounted for the largest share of the acquisitions, with a total of 30,718 acres, for $537,368, or $17.49 per acre in nominal terms, or $1,990,252 and $64.78 per acre in 1967 prices.[50] After 1905, when the word was out that Los Angeles was buying land to get water, the negotiations for other properties appear to have gone smoothly and quickly. There is no published or archival record of disputes over price. These negotiations involved the least productive farms in the region with the fewest long-term prospects and lowest reservation prices. By 1921, of the total land acquired by Los Angeles, 72,400 acres were in the Independence region and 22,600 in the Bishop area, and most were riparian to either Owens River or Bishop Creek Ditch.[51]

There was a lull in acquisition activity from that year through 1922. Then in 1923, activity increased as the LADWP began to buy properties in the northern Owens Valley. In this part of the valley there were more farms, more valuable lands, and more people living in the region's five small towns. The period 1923–34 is when most of the land and water rights transactions took place, and these transactions are the source of the real Owens Valley controversy—the battle over price. While many of the negotiations went smoothly, others involved disputes over property valuation and what Los Angeles was willing to pay. The bargaining disputes spilled out of the valley and into the local and national press, the mayor's office in Los Angeles, and the governor's office and the state legislature in Sacramento. These negotiations are the focus of the analysis presented here because they explain why Owens Valley has achieved such notoriety and why there are lessons from this experience for today's water markets.

During the lull in property acquisitions between 1911 and 1922, there were still some disputes between the LADWP and landowners (particularly with Fred Eaton) and power companies over the location and size of a potential water storage reservoir in Long Valley, how much land could be maintained in irrigation with it, and development of power sites. The uncertainties of water demand in Los Angeles, given its spiraling growth, and the residual amount to be made available for Owens Valley agriculture affected estimates of the size of the reservoir

(a)

(b)

Figure 3.4   (a) Putting Concrete Cover on Conduit in Mojave Division (May 1909); (b) Concrete-lined Conduit, Mojave Division; (c) Lining the Conduit in Mojave Division; (d) Opening the Headgates of the L.A. Aqueduct.

SOURCE: J. B. Lippincott Collection; LIPP 142: 650C. Water Resources Center Archives, University of California, Berkeley.

that would have to be constructed. A relatively low dam and smaller reservoir could have been built on land owned by Los Angeles. But a higher dam and larger reservoir to meet new demand projections required additional land from Eaton, who held out, demanding over $1 million for it. Ultimately, no agreement could be reached and the city abandoned the reservoir project.[52]

(c)

(d)

Historical accounts of the Owens Valley controversy emphasize the importance of the lack of a sufficiently large reservoir that might have sustained irrigated agriculture in Owens Valley while supplying Los Angeles with water, thereby avoiding subsequent conflicts between the city and landowners over land and water rights.[53] This is unlikely to have been the outcome, however. Los Angeles eventually required all of the water rights in Owens Valley, and most farmers wanted to sell for the most favorable prices they could get. Conflict over the terms of trade would not have been avoided by an even larger reservoir site.

As it was, most farmers in Owens Valley through 1923 were not much affected by Los Angeles's early purchases of land, and they benefited from increased

TABLE 3.3

Acquisition of Owens Valley properties by Los Angeles, 1905–34

Year	Number of parcels	Acres[a]	Purchase price (dollars)[b]	Price/acre (dollars)[b]
1905	5	20,298	1,707,733	84.13
1906	36	27,043	1,016,530	37.59
1907	39	12,057	727,761	60.36
1908	39	6,218	396,619	63.79
1909	40	9,668	791,232	81.84
1910	7	1,090	50,821	46.63
1911	7	2,120	5,686	2.68
1912	11	10,736	101,969	9.50
1913	2	103	8,653	84.01
1914	0	0	0	0.00
1915	2	240	50,000	208.33
1916	0	0	0	0.00
1917	1	240	4,167	17.36
1918	2	250	554	2.22
1919	5	2	5,888	2,944.02
1920	3	840	50,333	59.92
1921	8	594	20,479	34.48
1922	2	520	98,606	189.63
1923	84	19,170	4,207,479	219.48
1924	137	28,992	3,554,783	122.61
1925	43	4,030	813,232	201.79
1926	149	14,399	5,804,526	403.12
1927	291	38,450	12,386,319	322.14
1928	14	1,642	183,333	111.65
1929	48	8,707	1,500,415	172.32
1930	29	4,902	560,340	114.31
1931	130	22,877	5,299,105	231.63
1932	17	2,172	256,897	118.28
1933	13	15,897	2,825,871	177.76
1934	3	8,845	1,659,352	187.60
Total	1,167	262,102	$44,088,686	$168.21

SOURCES: Various and often slightly conflicting sources were pieced together for this record. Consistency and completeness were the primary criteria for selecting sources for the table. The differences are generally small, with entries in one year presented for next year a principal difference:

For the years 1905–14 and 1916–21, the data are from Charles H. Lee, "Statement of Lands in Inyo and Mono Counties Owned by Department of Public Service of the City of Los Angeles Prior to January 1923," January 1924, Lee Folder, MS 7611 98L4a, Water Resources Research Center Archives, U.C. Berkeley. For 1915: memo from D. L. Hirsch to A. J. Ford, August 2, 1927, "Report of Land Purchase in Owens Valley from 1905 to March 1925," Owens Valley Lands file, Tape GX0007, LADWP Archives. For 1922–24: "Land Purchases in Inyo and Mono Counties to June 30, 1929, Miscellaneous file, Tape GX0004, LADWP Archives. For 1925–34: "Tabulation Showing Status of Ranch Land Purchases Made by the City of Los Angeles in the Owens River Drainage Area from 1916 to April 1934," prepared in Right of Way and Land Division by Clarence S. Hill, Right of Way and Land Agency, compiled by E. H. Porter, April 16, 1934, Tape GX0004, LADWP Archives.

[a] The different sources do not always agree on the total land purchases by year.

[b] All values are in 1967 dollars. Deflator, University of Michigan Library Historical Consumer Price Index, 1880–1998 (http://www.lib.umich.edu/govdocs/historiccpi.html).

demand for their farm products from the workers involved in the construction and maintenance of the aqueduct, as well as from high commodity prices associated with the buildup to America's entry into World War I. Even when the agricultural economy began to deteriorate elsewhere in the United States in 1921, conditions remained generally prosperous in Owens Valley communities.[54] A new highway connected the region to Los Angeles's burgeoning population, bringing greater flows of visitors to the valley, as it continues to today. All of these activities brought additional commercial sales, at a time when the old Nevada mines that had previously been the source of local demand were dwindling. All in all, this was a time of relatively good relations between the residents of Owens Valley and the LADWP.[55]

LOS ANGELES BUYS THE LAND AND WATER RIGHTS IN THE VALLEY, 1923–35

The flow in the aqueduct between 1913 and 1922 was thought to be sufficient for Los Angeles to rely on surplus water and the maintenance of perhaps 30,000 acres in irrigated agriculture, primarily in the upper part of Owens Valley around Bishop. There probably had never been more than 35,000 acres of cultivation in the valley, although Conkling estimated that perhaps double that figure might have been possible with sufficient drainage.[56] Even so, given the limited agricultural potential of the valley due to its elevation, remoteness, terrain, and soil quality (like much of the rest of the Great Basin), it is unlikely that such expansion would have been economically feasible.

Beginning in 1923, in the face of rising population growth in Los Angeles (up 81 percent in ten years to 576,673 people by 1920) and the onset of a severe drought that lasted through 1926, the LADWP began to purchase land in the more agricultural and densely populated upper part of Owens Valley.[57] The city was now dependent upon the aqueduct and had to ensure that its flow was augmented. The drought demonstrated that its holdings in Owens Valley and claims on residual water in 1923 were insufficient to fully support the aqueduct during dry years. An independent engineering consultant, Thomas Means, issued a report in 1924 calling for increased groundwater pumping and extensive purchase of lands for their surface water rights.[58]

As the data in Table 3.3 reveal, the LADWP began to dramatically increase its purchases of land and water rights in 1923. Once the LADWP bought a farm, its ditch water allocation, riparian claim, or groundwater could be released to flow down the river to the aqueduct. Depending on aqueduct needs, some water was retained for more limited irrigation on the newly acquired farm, and livestock raising replaced other kinds of agriculture. The farms were consolidated into larger units and leased by the LADWP, a practice that continues today. If agency land agents could not reach a sales agreement with one landowner, they would turn to another. The LADWP purchased properties more or less consistently until halting new purchases between May 1927 and January 1929.[59] It then resumed buying properties from 1934 through 1945, when there was little left to buy that offered any water. It also began negotiations to purchase town properties.

Drought substantially reduced the aqueduct flow, which fell to 50 percent of its 1922 level in 1924 (see Table 3.2). At that time 71 percent of the aqueduct's water was directed toward urban use, reflecting the increased pressure on the LADWP to secure additional supplies. It had a smaller cushion of San Fernando Valley agriculture water use from which to draw. Through purchases of additional land and water rights in Owens Valley and groundwater pumping, aqueduct levels rebounded, reaching a new high in 1927. By that time, however, the LADWP was in search of even more water supplies and considering expansion north into the Mono Basin. The city began purchasing property there in 1930.

As the LADWP acquired properties in order to release their surface water to the aqueduct, it also began to pump groundwater to secure additional supplies. This practice, however, lowered water tables during the drought, affecting adjacent landowners, whose wells began to go dry. Under California law, parties who could demonstrate harm from groundwater withdrawal could block pumping on city lands. Under a State Supreme Court ruling in *San Bernardino v. Riverside* (186 Calif. 7, 15, 1921), groundwater could not be pumped and removed if doing so would damage surrounding lands. In these cases, Los Angeles moved quickly whenever possible to purchase those properties in order to allow pumping to continue and maintain aqueduct flow.[60] In addition to any legitimate claims of damage, the court ruling gave all nearby farmers the opportunity to strategically threaten the city with court injunctions unless their properties were purchased at a price favorable to them. Disputes and litigation over pumping continued to plague Los Angeles throughout the period 1924–35, with farmers asserting harm and demanding a halt to pumping, compensation, and the purchase of their properties. The agency was vulnerable to these claims since, although the amount of damage was often difficult to verify, local juries were expected to favor the claimants. In the meantime, the water needed for the aqueduct during drought was tied up. To internalize any potential third-party effects, purchase was usually the most expedient option.[61]

By the 1920s, whenever Los Angeles was buying properties for their surface water rights or access to groundwater, it was certainly no secret who was buying the land or why. Negotiations over some sales were contentious and protracted, taking from three to five years, particularly for the properties with the most water. For those farmers with drier, desert lands and less water, however, agreements generally came smoothly. Farmers often wrote the board offering to sell their properties. In 1925, board counsel W. B. Mathews commented on the "insistent demand" by some property owners for the city to buy their lands.[62] The board reported that "the prices paid, with few exceptions, have been entirely satisfactory to the seller."[63]

The Ditches of Owens Valley

The lands of primary interest to the LADWP were those that carried the most water and were either properties riparian to the Owens River and some feeder streams or more important, organized as part of formal irrigation ditch companies. There were at least fifteen ditches or canals used for irrigation in an otherwise arid valley. In total, there were 110 miles of primary and secondary ditches lacing the north valley, where most agriculture took place.[64]

The construction of ditches required cooperative investments, so farmers joined to incorporate mutual ditch companies and to place joint appropriative water claims. Member farmers held stock in the company, and expenses were met through annual assessments. The ditch companies usually held the water rights, with their priority based on the date the company was organized. Water was apportioned among farmers in proportion to the number of shares each held in the company.[65] Earlier ditches had higher-priority claims than newer ditches, although given the vast amount of water available in Owens Valley, priority issues were not important except during drought. Further, the location of a ditch's intake from the Owens River also granted strategic advantages: ditches farther up the valley could divert water before their downstream neighbors could do so, a practice shortly learned by Los Angeles.

Table 3.4 provides summary information for eleven of the major ditches in Owens Valley, including the McNally Ditch, Bishop Creek Ditch, the Owens River Canal, the Owens River and Big Pine Canal, the Rawson Ditch, and Farmers Ditch.[66]

A number of points can be made using the data in the table. Approximately 53,000 acres of land were associated with these ditches, 38 percent of all the land in farms listed in the 1920 Census for Inyo County; they covered the majority of

TABLE 3.4

Leading ditch companies in Owens Valley, 1923

Ditch company name	Incorporation (appropriation) date	Acres	Appropriation water claim (inches)	Mean diversion water (inches)	Diversion fraction	Assessed value of ditch property (dollars)
Owens River Canal	1886	8,174	5,000	3,050	.610	809,850
Russell	1886, 1889	910	2,000	400	.200	83,000
Bishop Creek Ditch	1878, 1893	10,433	6,000	4,350	.725	1,168,225
Love	1874, 1880, 1888	830	2,000	400	.200	62,250
McNally	1877	10,539	7,000	4,150	.593	1,065,775
Farmers	1888	2,140	2,000	900	.450	176,700
Geo. Collins	1877	1,057	2,000	450	.225	101,825
Rawson	1886, 1887	8,352	4,000	1,300	.325	579,250
A.O. Collins	1881, 1886	855	708	300	.425	69,675
Owens River and Big Pine Canal	1888	6,714	5,000	2,000	.400	636,100
Sanger	1886	2,500	2,000	600	.300	147,000
Total		52,504	37,708	17,900	.470	4,899,650

SOURCE: Charles H. Lee and S. B. Robinson, "Use of Water for Irrigation in Owens Valley in Connection with the Supply of the Los Angeles Aqueduct," Table I for incorporation and appropriation dates, and other data from "Recent Purchases of Water in Owens Valley by City of Los Angeles," Tables I, II, November 1923, Cope Rand Means Co. Engineers, San Francisco, Lee Folder, MS 7611 98246, Water Resources Research Center Archives, U.C. Berkeley.

the approximately 40,000 improved farm acres also listed in the census, as well as much of the roughly 30,000 acres in irrigation.[67] Unless a farm was along a ditch or riparian to the Owens River or a dependable feeder stream, of which there were few other than Bishop Creek in Owens Valley, sustained agriculture was not possible in the arid region, except through groundwater pumping. But there probably were no farms that relied solely on groundwater at this time. Ditches and the river crisscrossed all of the relatively flat, low-elevation land suitable for farming. Any farm that depended on pumping would have also been in comparatively inhospitable terrain. Given the small size of the region's farms on average, these farms would have been extremely minor.

On average, the ditches diverted only about 47 percent of the water they claimed under the appropriative doctrine. In part, this reflects some seasonality and the variability of water supplies in the region. During drought periods, the ditches would have diverted more, perhaps all, of their water.[68] This circumstance also reflects the large amount of water in the region relative to available arable land. There just was not enough good, flat land for irrigated agriculture. The scarcity of arable land also meant that there were no other water claimants who could assert that the ditches were not placing their water fully in beneficial use. Such individuals might have competitively claimed the excess water (had they been able to survive socially in the small, isolated communities in Owens Valley), but they would have had little or no land to devote it to and hence would also have violated the beneficial use mandate.

Los Angeles also might have purchased just the excess water, and in a sense that is just what the city did before 1923, relying on the residual flow in the Owens River to fill the aqueduct. But after 1923, with growing urban demand in the face of drought, this practice was no longer possible. In 1923 the board's objective was to acquire sufficient properties to raise its water supply from the aqueduct from approximately 290 cfs of water to 460 cfs of water per day (209,953 and 330,030 acre-feet per year, respectively).[69]

In 1923 the estimated mean flow of the Owens River at Bishop at the entrance of the valley was 449,000 acre-feet. Of this, 234,000 acre-feet were diverted to agriculture by the various ditches and riparian users, leaving 215,000 acre-feet in the river. With seepage from irrigation, the outflow from the valley was estimated at 295,000 acre-feet, or 408 cfs.[70] The Owens River flow fluctuated, depending on precipitation levels, and the LADWP sought additional supplies in face of unprecedented population growth in the city, the uncertainty of drought, and a 1924 report by the Board of Engineers that predicted a water deficit in the city unless new supplies were forthcoming.

A 1924 report by Louis C. Hill, J. B. Lippincott, and A. L. Sonderegger depicted a combined water supply for Los Angeles of just 432,000 acre-feet per year from the aqueduct, runoff from San Fernando Valley irrigation (also Owens Valley water), the Los Angeles River, and Los Angeles Basin pumping, as compared with total urban, industrial, and agricultural demand of 564,500 acre-feet, for a deficit of 132,500 acre-feet.[71] The report outlined various contingencies for getting more Owens Valley water to meet the deficit in Los Angeles while at the same time sustaining 30,000 acres of land in irrigation. But by the end of 1924, the notion that irrigation could be maintained on so much land while still meeting the aqueduct's needs for water were dropped.[72]

The extra water not used by the ditches, as illustrated in Table 3.4, also raises two other points that are important for the Owens Valley story. One is the subsequent diversion by farmers on downstream ditches of the water Los Angeles obtained upstream and released to the river for the aqueduct. As described below, this practice was clearly a bargaining tactic. Although the farmers claimed that released water was "unused" under the beneficial use doctrine and therefore available for appropriation by other claimants, the data in the table show that farmers most likely did not need this water. They were not using what they already had. Second, the problem of how to value or "price" this excess water became the center of dispute between the city and farmers as described in Chapter Four. Since it was not generally used in agriculture, its local value was low, whereas its value in Los Angeles was high. Determining what price to set and who got the surplus, the farmers or Los Angeles, was a major source of the Owens Valley conflict.

In early 1923 the LADWP advertised in local Owens Valley papers that it was seeking to buy all properties on the east side of the Owens River and along the Bishop Creek Canal, except the Owens River Canal.[73] The Owens River Canal was the hotbed of organized farmer efforts to secure higher prices from Los Angeles, led by Wilfred W. Watterson and Karl Keough, members of the canal governing board. The McNally Ditch was of particular interest to the city because of its high priority (early) and large appropriation claim to water. By the end of the year, the LADWP had purchased forty-one individual properties and made one group purchase of twenty-five additional ranches, for a total of 15,500 acres, including nearly all of the farms under the McNally, Russell, and Big Pine Ditches and smaller parts of the Love, Farmers, A. O. Collins, George Collins, Rawson, Owens River, and Bishop Creek Canals. As indicated in Table 3.4, these were among the most important canals in the valley.

The thirty McNally Ditch properties cost approximately $1 million, and the twenty-five Big Pine Ditch properties cost another $1.1 million.[74] This translated to \$171.84 per acre for McNally farms and \$239 per acre for Big Pine farms, well above the usual per acre prices paid for land during 1923–24 as shown in Table 3.3. The prices paid for properties along these ditches and the speed of agreement with the LADWP demonstrated its determination to secure these properties and its dependence on Owens Valley water.[75]

The LADWP purchased the farms along the McNally Ditch first, and when it released their water to the Owens River for use in the aqueduct, farmers on the Big Pine Canal downriver diverted the water before it could reach the intake. Although the water claims of Big Pine Canal were inferior to (newer than) McNally's, its members claimed that since its water was no longer being deployed beneficially in agriculture, it was free for diversion and use by others. The LADWP countered with court injunctions, but given fear of delays, uncertain court action, and the need to secure the water flow, it bought the Big Pine farms within six months at prices higher than it had paid earlier for other ditch properties.[76] This clearly is evidence of successful holdup by the Big Pine farmers. As we will see, this was one of a variety of strategies used by the farmers to improve their bargaining position with the LADWP.

The Owens Valley Irrigation District

The 1923 acquisitions of the McNally, Big Pine, Russell, and other ditch properties on the east side of the Owens River and along the Bishop Creek Canal were motivated both by the amount of water involved (33 percent or more of the Owens River flow) and perhaps more important, by a desire to thwart the formation of the Owens Valley Irrigation District (OVID).[77] The farmers in Owens Valley were at a disadvantage in negotiating with the LADWP because of their large numbers and heterogeneous production potential and interests. Although there were various organizations representing landowners, including the Owens Valley Protective Association and Associated Ditches, to strengthen their bargaining position, by 1922 farmers in the eleven major irrigation ditch companies began the formation of an encompassing irrigation district to be the single negotiating unit with the LADWP.

The Owens Valley Irrigation District was organized under the Wright Act of 1887, which allowed for the formation of quasi-governmental organizations for irrigation purposes with the ability to issue bonds. The district was to be "a central body which might more effectively deal with the city of Los Angeles, which city has been securing the surplus water from the valley for the past twelve years."[78] The district was approved by the State Engineer and Inyo County

Board of Supervisors, in November 1922.[79] It was also agreed to by the governing boards of the ditches and an affirmative vote by the residents of 599 to 25 on December 26, 1922, and formally organized on January 16, 1923. It covered all of the arable land, some 54,000 acres, in the northern end of the valley, between the Owens River Canal in the west and the McNally Ditch in the east. Included in the district were 323 farms with an average farm size of 161 acres.[80] The companies joining the district were the Owens River Canal and Russell, Bishop Creek, Love, McNally, Farmers, George Collins, Rawson, A. O. Collins, Big Pine, and Sanger Ditches. Its members drew 156,000 acre-feet of water on average per year and had claims to a total of 180,000 acre-feet, over three-fourths of all the water devoted to agriculture in the valley.[81]

According to the plan, the water rights held for farmers in each of the ditch companies were to be transferred to the Owens Valley Irrigation District. The district board then would bargain directly with the Los Angeles Board of Water and Power Commissioners. Under the new irrigation district, the negotiating environment would have shifted from one of hundreds of farmers bargaining with the LADWP to one of two roughly equal entities. Figure 3.5 shows the area to have been included in the irrigation district.

In May 1923 the California Bond Certification Board authorized the issuing of $1.65 million in bonds by the Owens Valley Irrigation District.[82] Almost all of the money was to be directed not to infrastructure investment, but rather to the purchase of the water rights held by farmers in member ditches.[83] But final formation of the irrigation district collapsed with the defection of key property owners: William Symons, president of the McNally Ditch Company; George Watterson, secretary of the Bishop Creek Ditch; and Leicester C. Hall, treasurer of the Owens River Canal. These three had properties on the McNally Ditch, and they led most of the other property owners to sell their farms to the LADWP. Supporters of the irrigation district charged that the McNally Ditch leaders were bought off with 5 percent commissions on land sales to the city.[84] As shown in the statistical analysis in Chapter Five, farmers on the McNally and Big Pine Ditches (properties purchased shortly after those on McNally) earned substantial premiums for their lands.[85]

Once Los Angeles had the majority of the land and membership of the boards of directors of these ditches, it withdrew them from the Owens Valley Irrigation District. In addition, the LADWP successfully sued to block further issuance of OVID bonds, and only $470,000 in bonds was sold.[86] By the end of March 1925, the Owens Valley Irrigation District was effectively sidelined, and farmers negotiated either individually with the LADWP or through three sellers' pools

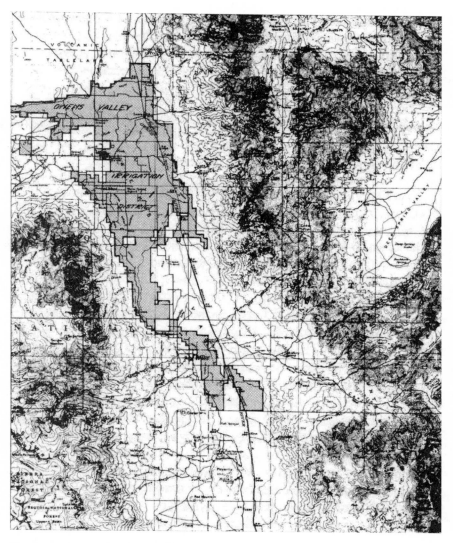

Figure 3.5 Owens Valley Irrigation District.

SOURCE: Los Angeles Department of Water and Power.

formed by adjacent farms on three important ditches.[87] As described in Chapter Four, these pools were successful in raising the prices received by their members for farmland, relative to the prices received by nonpool farmers.[88]

If the Owens Valley Irrigation District had been successful, a clear bilateral monopoly situation would have developed, certainly a better situation for the farmers in bargaining with the city. The efforts of farmers to form both the Owens Valley Irrigation District and then, later, three sellers' pools reveals a level of sophistication and commercial shrewdness that is not consistent with the usual portrayal of Owens Valley farmers as hapless victims of Los Angeles's ruthless land and water acquisitions.

Bargaining Conflicts Between Landowners and the LADWP

From 1924 through the rest of the decade, the LADWP bargained with individual farmers and with the sellers' pools to secure lands and water rights, especially along the remaining ditches, eventually obtaining 96 percent of the farmland in Owens Valley. Once the LADWP obtained two-thirds of the stock of a mutual ditch company, it could dissolve the company since extensive irrigation was no longer a priority.

In the meantime, negotiations for key properties with the most water were contentious as the owners and representatives of the LADWP maneuvered to secure the best prices. Farmers wanted land and water prices based on what they perceived to be values in Los Angeles, whereas the LADWP wanted to hold to local agricultural values. When bargaining impasses emerged, the agency's land agents would move on to negotiate with another farmer.

Holdouts could receive higher prices, but if the agency concluded that the negotiations were going nowhere, that there were no longer bond funds available to complete further purchases, or that the city had sufficient water rights and no more were required, then these holdouts could bear additional costs. Until the end of the decade there was also uncertainty as to just how much land and water Los Angeles would have to buy, given unexpected population growth and recurring droughts in the region. As early as 1911 the *Annual Report of the Los Angeles Board of Public Service Commissioners* stated, "What the ultimate needs of the City will be is impossible to foresee."[89]

These conditions affected the bargaining positions of both parties because landowners could not fully anticipate whether there would be additional bond issues and another round of land purchases, should their current negotiations fail. Given the potential gains from sale if prices were based on Los Angeles water values, landowners did not want to be left out of the potential bonanza. At the same time, LADWP officials did not know how much irrigated land they could leave if bargaining reached a stalemate.

Besides the opportunity cost of the lost sale revenue, holdouts could be left isolated on a ditch with higher ditch maintenance costs, since the city might not have contributed once its ditch water was released from its properties for the aqueduct. Weeds and insects from no-longer-cultivated properties also could infect remaining neighboring farms.[90]

Just how important these costs were is impossible to know given that the claims were made in the midst of a major public relations battle with Los Angeles that was part of the farmers' bargaining strategy. The statistical analysis reported in Chapter Five shows that farmers earned more by holding out. Whether this was sufficient ex post to offset any additional costs they might have borne is unknown, but the magnitudes of the gains are such that farmers likely were better off if they successfully held out.

It also was alleged by some farmers that the LADWP engaged in a "checkerboarding" strategy, buying properties around holdouts to force them to sell.[91] Since checkerboarding would be the natural outcome of the successful purchase of one property and the breakdown of negotiations on the next, it is difficult to separate what might have been purposeful and what was the result of an ongoing series of negotiations. There is no evidence of major checkerboarding or that this strategy successfully made holdouts worse off.[92] There are no pricing data suggesting that those who resisted sale were penalized. Rather, farmers who delayed sale to the city earned more, all things equal. Moreover, throughout the negotiating period, the LADWP tried to stick to a pricing strategy of paying the same prices for comparable, adjacent properties.[93] Other farmers, usually those on less favorably endowed properties with less water, were anxious to sell and were concerned about a lack of response from the LADWP to their sale offers.[94]

In addition, as negotiations among the LADWP, farmers, and town lot owners dragged on, there was growing concern about the safety of the water supply upon which Los Angeles depended. Between 1924 and 1931 the aqueduct and city wells were repeatedly dynamited, although the aqueduct was never seriously damaged.[95] The threat to the city's water was clear, as indicated in a letter circulated among Owens Valley ranchers on December 24, 1924:

> Something is surely going to happen before long. To avoid this city must pay for the ruined businesses what is demanded, this represents about fifteen million dollars, and in my opinion the city would do well to pay it and rid itself of the menacing situation. . . . This is nothing compared to the potential damage.
> I am amazed at the apathy of the city with reference to its water supply, evidently those in control do not realize the danger this big ditch is in, if something is not soon done to pacify the crowd here Los Angeles will face a disaster of the first magnitude.[96]

The most celebrated event in terms of press coverage was the November 16, 1924, seizure and opening of the Alabama Hills waste gates on the aqueduct, which released the water flow into the desert. Although the gates subsequently were closed, this act of sabotage was viewed favorably in the media and considered a public relations coup for farmers who were trying to attract attention to their cause from the broader citizenry.[97] Indeed, on December 26, 1924, the state engineer, W. F. McClure, issued a report to the governor regarding the opening of the Alabama Hills gates, siding with valley residents in their conflict with the Los Angeles Public Service Commission. He justified their actions, outlined their history of grievances against the city, and labeled the opening of the gates and the waste of aqueduct water as "a most popular move." McClure emphasized that either the valley be allowed to "go its own way . . . or that the [board] . . . be willing to pay well for the privilege of exercising [its domination of the valley]."[98]

These episodes of periodic violence, labeled "California's little Civil War" by the press, attracted state and national attention and helped compel the LADWP to reach agreement with property owners on price.[99] The agency viewed the dynamiting as a negotiating tactic, but was extremely worried about disrupting the aqueduct flow and maintaining an adequate water supply for Los Angeles.[100]

Throughout the 1920s, the press was invariably critical of Los Angeles, portraying Owens Valley farmers as battling a giant city against uneven odds, and losing. For example, in late March 1927, the Owens Valley Protective Association ran a succession of advertisements directed to the California governor and state legislature in the *Sacramento Union* that included "A Message from Owens Valley—The Valley of Broken Hearts" and "We, the Farming Communities of Owens Valley, Being About to Die, Salute You." And from March 29 through April 3 the paper ran a series of sensational articles by Fredrick Faulkner on the alleged destruction of Owens Valley, with titles like, "Owens Valley Plight Pitiful," "Land Grabbers Wreck Valley," "Flag Furled on Desolation," "Water Sharks Wreck Valley," and "Blooms Fade before Waste."[101] Other papers, such as the *San Francisco Call*, the *Los Angeles Record*, and naturally the *Owens Valley Herald*, also were extremely critical of the LADWP in its dealings with the farmers of Owens Valley. Even the *Los Angeles Times*, on November 18, 1924, called for justice for Owens Valley residents. All of this suggests that the farmers had been able to persuade others of the righteousness of their cause.[102]

Although much of the Owens Valley record involves conflicts over price between the city and landowners, there is evidence of concern among Los Angeles tax and water ratepayers that the city was paying *too much* for Owens Valley land. For example, land buyer John Merrill asserted in 1927 that while the city had paid an average of $200 per acre for Owens Valley lands thus far, the properties could have been secured for $50 to $75 per acre, for a total expenditure of $5 million rather than $12 million.[103] The *Hollywood Daily Citizen* ran an editorial objecting to any payment for town properties beyond appraised values.[104] In 1927, the Municipal League charged that landowners in Owens Valley were conspiring to "fleece" city taxpayers.[105] These allegations were of concern to the LADWP, whose reputation for sound management was at stake. The board's credibility affected its standing with the mayor and City Council as well as its ability to win citizen support for new bond issues to fund Owens Valley operations.[106]

The Board of Water and Power Commissioners was charged with providing a sufficient water supply for Los Angeles at acceptable rates, subject to the constraints of bond and rate revenues. The five board members were appointed by the mayor of Los Angeles for staggered-year terms, and the board was not always in agreement on how to deal with Owens Valley landowners.[107] There was particular turnover within the board between 1929 and 1931, when its members argued over how to resolve the longstanding controversy in the valley over the purchase of land and water rights.[108]

The mayor, in turn, had broader concerns, including maintaining the confidence of ratepayers, taxpayers, and voters and of key constituencies, such as the business community as represented by the Chamber of Commerce, the Municipal League, and the Clearinghouse Association.[109] The latter sought to promote the growth of Los Angeles and the minimization of negative publicity that could damage the city's prospects. Private water and power companies also were often in contentious negotiations with the LADWP over the sale of their distribution systems and power sites, and they occasionally joined Owens Valley landowners in contesting the prices offered by the LADWP.[110] Landowners in the San Fernando Valley depended almost completely after 1913 on Owens Valley water, and not facing the same political constraints as the board, they sought independent agreements with the Owens Valley Irrigation District in 1924 to ensure water flow in the aqueduct. This option was quickly squelched by the board.[111] All in all, then, the board was under considerable political pressure to obtain and guarantee a secure water supply for the city in a cost-effective manner. This pressure certainly influenced the stands taken by the LADWP in its dealings with individual property owners.

Concerns about violence, threats to the water supply, and negative publicity for Los Angeles led the governor's office, the Municipal League, and the Los Angeles Chamber of Commerce to offer mediation in 1927.[112] The Chamber of Commerce sent in an investigating committee and the Los Angeles Clearinghouse Association offered to arbitrate the dispute over land and water values between Owens Valley property owners and the LADWP.[113]

Members of the California legislature also entered into the conflict. Those who represented Inyo County and other similar rural jurisdictions naturally intervened on behalf of the Owens Valley landowners. For example, in 1927, Assemblyman Dan Williams of China Camp, a small western Sierra community concerned about its own water supplies, called on Governor Clement Young to intercede in the price disputes between Owens Valley farmers and the LADWP with mandatory arbitration. The governor declined, but led a special legislative investigating committee to the valley. The committee subsequently drafted a resolution condemning Los Angeles for its "ruthless destruction" of the valley. The resolution passed the Assembly but failed in the State Senate.[114]

The broad appeal of this and other legislative campaigns against Los Angeles that went beyond only rural California representatives reflected the growing concern of northern California representatives about the rapid growth of southern California and the gradual shift in political power to that region. Politicians from northern California took advantage of the Owens Valley conflict to blacken the reputation of the upstart metropolis to the south. Beyond members of the legislature, the state's governors were drawn into the valley's disputes because they had the broadest constituencies in the state and tried to be responsive to as many as possible. Accordingly, while reluctant to strongly support either side, California's governors wanted the conflict resolved. They added to the political pressure on the LADWP to end the discord in the valley over Los Angeles's acquisition of its land and water.

THE PROBLEM OF THE TOWN LOTS

As Los Angeles purchased properties in Owens Valley and took them out of irrigated agriculture, there were complaints that this action was hurting the local economy and damaging property values within the five towns. The magnitudes of the effects were disputed, but the conflict over the impact on town lot owners in Owens Valley is an example of the broader contemporary problem of third-party effects of water transfers throughout the American West. As shown below, it is very difficult to separate the various factors that could influence property values, to determine what amount of compensation might be warranted, and to distinguish legitimate claims for compensation from rent-seeking.

Merchants claimed to have lost one-third of their commercial trade owing to the decline in the agricultural economy.[115] The general fall in U.S. agricultural commodity prices in the 1920s would also have hurt the community, but this effect was difficult to separate from the consequences of the LADWP's purchase of lands and export of water. For example, census data reveal that the value of Inyo County crops fell from $1,503,195 to $791,257 between 1920 and 1924, a drop of 47 percent. This decline, however, occurred before most of the property purchases in the valley by Los Angeles. Further, during that same five-year period, the number of farms in the valley fell only 7.5 percent, from 521 to 482, unlikely enough to account for the observed fall in the value of agricultural production.[116]

Other evidence also suggests only a slight decline in economic conditions in the valley as the decade of the 1920s proceeded. It is important to recall that this was a time of a deteriorating agricultural economy throughout the United States, with its effects seemingly much more severe elsewhere than in Owens Valley.[117] Post office receipts in Big Pine and Bishop declined 5 percent between 1921 and 1926, from $18,741 to $17,752; grade school attendance fell 10 percent from its peak in 1923 to 798 in 1926; and voter registration declined from a high of 3,028 in 1923 to 2,614 in 1926.[118] Automobile registration and bank deposits in the region rose, however.[119] The most effective means of evaluating the impact of land purchases and water exports is to compare the record of Inyo County and Bishop with comparable counties and communities elsewhere in the Great Basin, affected similarly by the agricultural depression of the 1920s. This comparison is presented in Chapter Five, where it is shown that both Bishop and Inyo County did significantly better than other similar communities.

Nevertheless, town property owners blamed the actions of the LADWP for the alleged falloff of commercial activity. The agency countered by pointing to the beneficial effects of its investments in the valley to develop water and power sites and the growth of recreational activities from Los Angeles. The Board of Water and Power Commissioners appointed a Special Owens Valley Committee to assess economic conditions in the little towns. The committee members traveled to the valley in November 1924. The committee determined that the town of Big Pine was most vulnerable to any effects of land acquisition by Los Angeles.[120] They studied how it might be assisted through the subdivision of nearby lands for home purchases in order to attract a larger population. The board also commissioned a study of how farms might be organized around Big Pine to better sustain its economy.[121]

After visiting Big Pine, the Special Owens Valley Committee traveled to the valley's largest town and Inyo County seat, Bishop. They met with local ditch leaders Wilfred Watterson and Karl Keough to discuss options for maintaining existing farms and irrigated agriculture around Bishop. The committee's offers, however, were rejected. Both Watterson and Keough demanded that the LADWP *purchase* the rest of the valley lands (farms and town properties), claiming that the city had already gone too far. Watterson offered the lands under the Owens Valley Irrigation District for $365 per acre, a price generally far higher than any paid thus far by the LADWP (see Table 3.3), and the offer was rejected.

The committee members, in turn, refused to consider the purchase of town properties as sought by Watterson and Keough, claiming that since the town lots carried no water, the LADWP had no authority under its charter to buy them. In their report back to the board, the Special Owens Valley Committee concluded, "Fear of what the City may do in the future, combined with the hope that large money payments will be made by the City if the owners stand together and show a bold front, are the actuating motives behind the present situation."[122] With this, the battle over the price of Owens Valley lands and town properties that began with the organization of the Owens Valley Irrigation District and the agency's efforts to undermine it in 1923 was renewed. The bitter competition over the gains from trade for land and water heated up, continuing through 1935.

In 1925 an "Owens Valley Reparations Committee" demanded either that the board pay $5.3 million in reparations to valley landowners for the loss in farm and town lot values or that the city purchase all the remaining properties in the valley for $12 million.[123] Later, the reparations demands were reduced to actual claims, but Los Angeles resisted taking any action.[124] Not only were the prices for town properties well above what the city had been paying for other lands, but they carried few or no water rights. Hence the board was uncertain that it had the legal authority under the city charter to purchase such lands, which did not "supply the City with an adequate supply of pure water."[125] Members of the board concluded that they would be personally liable if they made such payments.[126] Beyond this, there were intense disagreements over valuation of the town lots, given both the export of water and the deterioration in the national agricultural economy in the late 1920s.

Legislation enacted by the California legislature in 1925 at the behest of Inyo County and other rural legislators required cities compensate businesses and property owners for damages when water was taken from the drainage area.[127] "An Act Providing for and Relating to Damage Resulting from or Caused by the Acquisition of a Water Supply or Taking, Diverting, and Transporting of Water from a Watershed . . . by a Municipal Corporation" passed overwhelmingly with 78 percent yes votes in the Assembly and 88 percent in the state Senate. The greatest support came from legislators from northern California, who not only were more likely to represent rural areas, but also wanted to constrain the popu-

TABLE 3.5

Compensation demands by the Owens Valley and Big Pine
Reparations Association Committees, by category

Reparations association	Number of claims	Category	Amount (dollars)
Bishop (Owens Valley)	55	Business	464,923.48
	15	Occupation	34,573.42
	91	Personal property	163,417.56
	9	Professions	34,418.62
	247	Real property	1,555,040.00
	14	Retail trades	18,004.66
Total	431		2,270,377.74
Big Pine	12	Business	66,831.73
	36	Occupation	24,660.00
	1	Personal property	857.12
	1	Professions	8,333.33
	63	Real property	432,597.51
	12	Retail trades	7,134.00
Total	125		540,413.69

SOURCE: Owens Valley Reparations Association and Big Pine Reparations Association Claims, Reparations file, Tape GX 0004, LADWP Archives.

lation growth of southern California. Support from southern California legislators for the legislation was more tepid, with 50 percent of the region's legislators voting no or abstaining in the Assembly.[128]

The statute added pressure on the LADWP to buy the town properties or be faced with hard-to-measure-and-agree-upon reparations demands. Indeed, the Bishop and Big Pine Reparations Committees presented the LADWP with 556 property damage claims totaling $2,810,791.[129] Table 3.5 provides a summary of the damages sought by the Owens Valley (Bishop) and Big Pine Reparations Committees, with the majority of the claims for reductions in the value of real property. But there were also claims for loss values to businesses, professional occupations (dentists, lawyers, barbers), retail trades, and personal property.

Los Angeles officials resisted the purchase of town lots until there was a state Supreme Court ruling in 1929 that authorized the acquisition of town properties by the LADWP even if they provided no water to the city. Funds to buy town lots and remaining agricultural properties in Owens Valley required a special bond election for $38.8 million, which passed in 1930.[130]

After 1930, negotiations between the city and town lot owners remained rancorous, requiring various appraisals, offers, and counteroffers, as well as another round of intervention by members of the state legislature.[131] The agency was obligated to buy the properties, and town owners knew that. Hence they held out for the best prices they could get. The LADWP, however, was constrained by available bond funds and allegations that it was overpaying for land. Even so, by 1935 Los Angeles had secured about 85 percent of the town properties. More discussion of the negotiations between town property owners and the LADWP is provided in the following chapter.

CONCLUSION

By 1935, the LADWP had paid $23,727,786 to acquire the ranches and farms from some 1,178 private owners in Owens Valley covering 281,003 acres. It also acquired 1,300 town properties from 824 owners for another $5,798,780. With this, the aqueduct's water supply from Owens Valley was secured, and the LADWP began to move rapidly into the Mono Basin to the north, buying more water and land there.[132] The acquisition of Owens Valley properties took thirty years, and it had not been a simple process. The generally collegial relations that existed before 1923 between the agency and many of the valley's residents deteriorated. State and city politicians, the press, and various civic and commercial organizations became involved. The aqueduct and city wells were dynamited.

The Owens Valley controversy was launched, to be repeated, reinterpreted, and often misrepresented for the next seventy years. The experience influenced subsequent legislation regarding water rights purchases and the export of water in California and colored how rural-to-urban water transfers would be forever viewed. The standard interpretation is that the conflict was between outgunned, but tenacious small farmers and a growing city that was gulping their water and livelihood. The historical summary outlined in this chapter, however, suggests a different story.

There was a growing city, Los Angeles. And there were small farmers in Owens Valley with water and land to sell. Given the decline in the U.S. agricultural economy in the 1920s and 1930s, they likely knew that their farms were too small to be sustained in their present state over the long haul, especially given the deterioration of rural communities across the country. Accordingly, even had Los Angeles never been part of the Owens Valley picture, consolidation and outmigration would have taken place, just as was occurring in other Great Basin areas. Because the farms were much smaller in Owens Valley than elsewhere in the region, the process would have been even more dramatic. With the agricultural depression of the 1920s, this inescapable situation was facing farmers throughout the area. Los Angeles's demand for land and water rights offered small farmers a wonderful opportunity out of the growing dilemma they were facing.

As the 1920s progressed, it became clear that Los Angeles would have to buy all of the valley's farms and water rights to meet the needs of the aqueduct and the city that depended upon it. By buying up the properties, Los Angeles internalized many of the externalities associated with surface water export and groundwater pumping. These issues are addressed in more detail in Chapters Six and Seven.

But the issue at hand in the 1920s was how much the LADWP would pay for the farms. Would those prices be based on the value of water and land in Owens Valley agriculture, as the LADWP naturally desired, or would they be based on the value of water in Los Angeles as the farmers naturally sought? This battle over the gains from trade and how approximately 1,000 farmers could improve their bargaining position relative to the city is the story of Owens Valley. The bargaining underlying these property negotiations is examined in the next chapter.

4 THE BARGAINING COSTS OF LAND

AND WATER RIGHTS EXCHANGES

IN OWENS VALLEY

You by now have learned that you are the Unanimous choice for the Third
man in the settlement of the Farm Land values. I have wished so much that
this could be and now it is if you will only accept. I hope that you will. You
can be a great help to us and I think also the City. It will not be so difficult
as you might at first think for the people here to have had so many different
strokes that they will welcome "One Great Strong Man."

*Letter from farmer Fred R. Smith to Edward Goodenough, Arbitrator,
July 18, 1929*

The battle over Owens Valley is one of the signature events in western American water history, policy, and markets. It was a battle with drama and violence in which both the press and politicians played roles. There were unscrupulous winners, the Board of Water and Power Commissioners of the big city of Los Angeles, and admirable losers, the beleaguered small farmers of Owens Valley. So the story has all the elements to make it an attractive parable to be repeated and used by those involved one way or another in conflicts over western water today.

But Owens Valley was in many ways a commercial transaction, though this characterization will be rejected by some. It involved the buying and selling of land and water rights. In some ways it was not unique. There was a long tradition of land sales by settlers on the frontier by the early twentieth century. The claiming and subsequent selling of frontier land was a major way by which migrants acquired wealth, and there is no reason that Owens Valley residents would have been different.[1]

The question is, then, what about these transactions over land and water made them so difficult? Additional questions come to mind. Why did they take so long—three to five years or more for some farmers? What ultimately was the distribution of the gains from trade? That is, who won and who lost? Or did everyone gain something? And finally, what lessons might be learned from this tortuous history for contemporary water markets?

Fortunately, analysis of the detailed records of bargaining and the related data sets deposited in the Los Angeles Department of Water and Power Archives, the

Water Resources Archives at U.C. Berkeley, and at the Eastern California Museum in Independence, California, allows for answers to these questions and provides more clarity on the overall Owens Valley water transfer than has been presented in the past. From these documents it is possible to outline the bargaining positions of the parties, their strategies, and key issues of contention as they negotiated over land and water. In addition, there is a compilation of 869 farmland purchases, including year of purchase, amount paid, location, and owner, as well as other property characteristics and membership in sellers' pools.[2] These data are used in the econometric analysis. There also are data in the U.S. Census and the State Board of Equalization Annual Reports that help in assessing the nature and outcome of the trades.

The answer to the first question about what is different about Owens Valley is that there was a single buyer, a government agency, the LADWP. This is unusual, although contemporary water transactions may involve a single buyer, such as the Metropolitan Water District of Southern California, as well. In fact it was one of the largest private land and water acquisitions by any local government in American history. In the 1920s and 1930s, 1,167 farms covering 262,102 acres were purchased for $20,768,233 ($219,727,905 in 2003 dollars).

A single large buyer introduced the potential for monopsony power. The farmers, in turn, organized collectively to increase their bargaining power as monopoly sellers. Bilateral monopoly conditions emerged, and these are notoriously difficult to resolve amiably, as many historical labor conflicts between unions and management attest. This means, though, that there will be a great deal to sort out in the transactions between the LADWP and the farmers in terms of price and the gains from trade. Because the buyer was *Los Angeles*, the epitome of unplanned urban sprawl, the exact opposite of the idyllic caricature of resolute Owens Valley farmers described in the literature, those who have examined Owens Valley in the past have focused on the inequality of the negotiations and missed the bargaining savvy and commercial maneuvering of the farmers themselves.

The bargaining problem between the LADWP and organized farmers is an interesting one worth examining in more detail. Insights from the analysis explain why some Owens Valley negotiations were so drawn out and acrimonious, while others were not, as well as the basis for the contemporary notion of water theft. The analysis also reveals how the bargaining strategies of buyers and sellers who are competing over the gains from trade can have important effects on the transaction costs of exchange and the overall perception of fairness.

ECONOMIC THEORY

Exchange requires locating the relevant parties; measuring the attributes of the asset to be traded; negotiating a sale price; drafting contracts; and enforcement. The transaction-costs literature emphasizes that each of these activities can be complex, affecting the timing, extent, and nature of trade.[3] The bargaining setting also affects the costs of negotiation.[4]

If the transactions to buy water-bearing farmland had taken place in a nearly perfectly competitive market with multiple buyers and sellers, then competition among parties on both sides would have generated information about property characteristics and willingness to sell and buy. There would have been little room for opportunism by either party in negotiations in misrepresenting positions or strategic behavior. Competition would have resulted in market-clearing prices for each property in reasonably short order. The distribution of the gains of trade of moving water from Owens Valley to Los Angeles would have been molded by the relative elasticities of urban demand and rural supply of water. Stalemates in negotiations would have led to both buyers and sellers to look for alternatives. Although there would have been many transactions, aggregate measurement, negotiation, and compliance costs would have been comparatively low, even with heterogeneous land.[5] The Owens Valley transfer would have looked like other routine land transactions.

Similarly, if there had been either a monopsony (LADWP) or a monopoly (a single landowner or a functioning Owens Valley Irrigation District) and the other side had been close to perfectly competitive (many land and water buyers or many land sellers), then the transaction costs of exchange also would have been low. Final sale parameters would have been dictated by the side with market power. This side would have captured most of the gains from trade.

But the bargaining setting for the most important farms was one of bilateral monopoly between the LADWP and the sellers' pools formed after the failure of the Owens Valley Irrigation District. Bilateral monopolies have indeterminate pricing outcomes because they depend on the relative bargaining power of the parties. Each party has an incentive to misrepresent its position in order to extract a greater share of the gains of trade in such negotiations, and there is little competitive pressure to force the revelation of more accurate information. Accordingly, bilateral monopoly negotiations often break down and take a long time to complete.[6] The issues to be addressed in this case are the extent of market power achieved on each side and how these conditions affected the costs and timing of exchange. The data set allows for an examination of the relative bargaining strengths of the LADWP and the sellers' pools of farmers.

THE NEGOTIATING ENVIRONMENT

The LADWP and farmers negotiated within an agricultural land market in order to get (sell) water. The item traded was farmland that differed in agricultural productivity and in the amount of water it could provide. Since the farms were very small on average, whole units were sold, rather than parts of them. The bargaining positions of the two parties were determined by the constraints they faced and their market power.

The Board of Water and Power Commissioners

On one side was the Board of Water and Power Commissioners or LADWP, which was essentially the only buyer of farmland in Owens Valley.[7] As described in the previous chapter, the mayor appointed the five commissioners, and the agency was charged with providing a relatively constant, reliable flow of water per capita in the aqueduct. It was also under ratepayer oversight in the management of its funds. The Los Angeles Aqueduct was a large, fixed, immobile investment that depended on Owens Valley water. As the city's population grew, the board had to acquire more water-bearing land to maintain its targeted aqueduct flow. The board's land purchases were financed through water bond issues.[8] Each new bond issue required voter approval, and multiple bond issues were floated between 1905 and 1930 for Owens Valley purchases and water infrastructure. Not all bond elections were successful, however. At least two proposed bond issues, in 1917 and 1929, were defeated by Los Angeles voters.[9] For all of these reasons, the board sought to acquire water-bearing lands quickly to ensure water supplies, smoothly to minimize transaction costs, and cheaply to stay within its water bond limits.

Eventually, the board required virtually all of the water in Owens Valley. The total water provided by all 595 farms in the data set examined below was 266,429 acre-feet, just above the aqueduct flow recorded in 1927 of 265,231 acre-feet.[10] The LADWP was interested in the farms with the most water. Buying them brought water for the aqueduct through a relatively few transactions. In contrast, the agency would have to buy more farms with less water in order to achieve a targeted aqueduct flow. But the farmers who had the most water usually were also the ones who organized collusively to delay sale and extract more of the gains from trade.

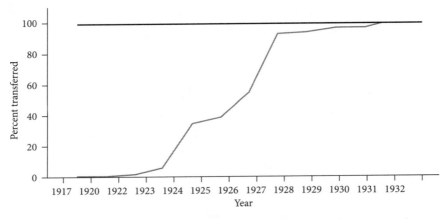

Figure 4.1 Percentage of Water from Owens Valley Transferred to Los Angeles, 1916–34.

SOURCE: "Tabulation Showing Status of Ranch Land Purchases Made by the City of Los Angeles in the Owens River Drainage Area from 1916 to April 1934," Prepared in Right of Way and Land Division by Clarence S. Hill, Right of Way and Land Agency, Compiled by E. H. Porter, April 16, 1934, Tape GX0004, LADWP Archives.

NOTE: The years shown on the horizontal axis are when there were observed transactions.

Figure 4.1 details the gradual acquisition of water rights in Owens Valley by showing the cumulative percentage of total water acquired at the same time each year. The figure covers the northern part of the valley where most of the farms were located and where most of the agriculture took place. Each observation in the data set includes information on the water available annually from a farm. The sum of these amounts represents the total water acquired by Los Angeles by 1934 and serves as the denominator of the fraction charted in the figure; the numerator is the cumulative water acquired by the city as of each purchase date.

As shown in the figure, there were major land (water) acquisitions in 1923, 1924, 1926, and 1927. Most of the delayed purchases of farms with more than the mean amount of water per acre, including those that remained unsold in 1927, were part of collusive organizations. By 1928 many of the water-bearing lands in the valley had been bought by the agency, and negotiations continued for the remaining properties. Had there been no bargaining conflicts and had sales been competitive, the pattern of water acquisitions shown in the figure would have shifted to the left between 1923 and 1934. Accordingly, the added costs of the observed transactions include the additional resources devoted to bargaining by both sides and the losses from delay in the acquisition of farms with the most water per acre and the postponed reallocation of their water to the city.

As shown in Figure 4.1, in 1923 when there were many farms available and considerable water remaining in the valley, the board could pick and choose properties to buy if the farmers were competitive in their sales efforts. If bargaining impasses ensued with some farmers, the board could move to other properties. After mid-decade, however, as the LADWP approached a supply constraint in Owens Valley, farmers' bargaining power increased, especially if they held significant amounts of residual water and were collectively organized.

In bargaining with farmers, the LADWP sought to buy farmland and the water rights associated with it based on their agricultural value in Owens Valley, rather than Los Angeles water values, which were much higher. If the LADWP could obtain the farms at these prices, the total surplus from reallocation of water would go to the city's landowners.

Since agricultural productivity endowments were known best by the farmers and less well known by agency personnel, the LADWP established an appraisal committee to collect data on each farm's characteristics. To reduce disputes over valuation, the LADWP selected a group that would be viewed as credible and acceptable to both parties. In 1925, a special Appraisal Committee was set up composed of "three of the leading citizens of Owens Valley": George W. Naylor, chair of the Board of Supervisors of Inyo County; V. L. Jones, Inyo assessor; and U. G. Clark, former county assessor.[11] Even so, since the Appraisal Committee was employed by the board, it was viewed with suspicion among landowners. During negotiations with some farmers in 1926, the credibility of the committee's prices was questioned: "You hired that committee; we had nothing to say about it . . . if you people hire these men, you expect them to go into the field and do as you tell them don't you?" Both the appraisals and the committee often were rejected: "They have been your committee for a long time. Let us forget them."[12]

During the valuation process, agency's land agents collected information about each farm—location, water rights, amount of irrigated land in cultivation, pasture, "brush" land, orchards, and improvements—and submitted the information to the Appraisal Committee. The committee compared this information with that for farms that had already been purchased to arrive at an "appraised value." The LADWP generally used a fixed multiple, usually 4.1 times appraisal value, to determine its offer price.[13] The agency wanted its offer prices to be based on "the fair average prices which the city had paid for substantially similar property in that region."[14] It repeatedly resisted adjusting prices above what it had offered for comparable lands in an area. This behavior is consistent with that of a monopsonist practicing price discrimination and moving up the supply curve of land.

For instance, in 1926 in asking the Appraisal Committee to determine offer values for properties under consideration, one of the agency's land agents stated: "It is also to be understood that these properties are to be appraised in the same manner and on the same basis that you have appraised other properties of substantially the same character and in accordance with previous values."[15] Available data do not include the estimated values of the productivity variables assembled for each farm to determine how effectively the agency kept to the rule. As described below, it appears to have been able to do so under some bargaining settings, whereas for others, it had to deviate from the rule. The data allow for tests of the farmers' market power, which could have forced departure from the pricing rule and increased final sale prices.

When the LADWP successfully reached agreement with a farmer, it acquired the right to the water associated with the land, either a riparian claim (less important in Owens Valley) or appropriative claims to ditch water and/or groundwater beneath the surface property. It could then release the farm's water from the ditch and/or pump groundwater for flow down the Owens River to the aqueduct intake. The farm typically was then consolidated into a larger unit for cattle ranching or less-water-intensive agriculture and leased back to one of the former owners.[16]

Farmers

On the other side were the farmers. Almost all farms were small, with each holding only a limited portion of the total water in Owens Valley. Hence no one farmer had market power, and in general, no single holding was critical to Los Angeles's water supply. In negotiations, farmers had to receive at least the present value of the agricultural productivity of their farms as their reservation price. They also sought as much of the gains from trade as they could secure. As a result, there would be bargaining disputes between the LADWP and farmers over both the accurate valuation and the distribution of the surplus. Any disputes that significantly delayed sale only would involve farmers with market power, however, since competitive farmers would reveal their reservation prices and eventually accept offers that met them.

There were two groups of farmers: those in drier portions of Owens Valley, and those on major irrigation ditches or along the Owens River, who had the most water and therefore were of most interest to the agency. Figure 4.2 plots the distribution of the 595 Owens Valley farms by water per acre along with the mean. Farms' water holdings varied considerably. The farms with more than the mean

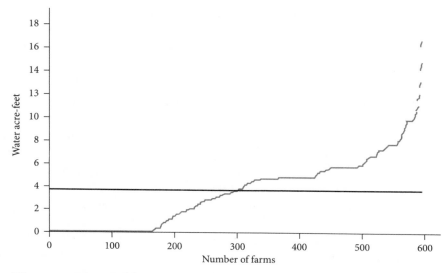

Figure 4.2   Water Holdings on Owens Valley Farms, 1916–34.

SOURCE: "Tabulation Showing Status of Ranch Land Purchases Made by the City of Los Angeles in the Owens River Drainage Area from 1916 to April 1934," Prepared in Right of Way and Land Division by Clarence S. Hill, Right of Way and Land Agency, Compiled by E. H. Porter, April 16, 1934, Tape GX0004, LADWP Archives.

water per acre were on ditches, whereas farmers with less water were distributed throughout the valley, away from major water sources. Being on ditches as part of mutual ditch companies provided a ready mechanism for the farmers with the most water to organize a sellers' pool to jointly bargain with the LADWP. We now turn to an examination of their organization and effectiveness.

The Sellers' Pools

Three sellers' pools were formed by owners of small clusters of adjacent, relatively homogeneous farms on two important ditches in 1923 and 1924. These farmers controlled about 17 percent of the valley's water.[17] Not all farmers on these ditches joined one of the pools, but most did. The organizations included the Keough pool on the Owens River Canal with twenty-three members, the Watterson pool of twenty members on Bishop Creek Ditch, and the Cashbaugh pool of forty-three members also on Bishop Creek Ditch. The pools were organized and led by the largest landowner in each cluster, who acted as the bargaining agent for all pool members.[18]

In their negotiations with the LADWP, pool leaders coordinated their stands through organizations such as the Owens Valley Protective Association in joint efforts to pressure the LADWP to meet their price demands. They also were involved in occasional violence against the aqueduct, and they led appeals to state politicians and the press. Negotiations between the board and members of these pools took the longest, some lasting through 1935 before final agreement was reached. They were the center of the bargaining disputes that plagued Owens Valley. These are the negotiations that made Owens Valley famous.

Pool members potentially had more bargaining power, and they held out for more of the surplus by delaying sale of their properties. By postponing sale, the farmers could wait until the LADWP was confronted with fewer options and thereby be able to extract higher prices. For the pools to be effective, however, they had to retain their members and avoid defection.[19] If only a few small farmers left a pool, it could retain its effectiveness for those that remained. But the defection of a large farmer, especially the pool leader, was a serious blow. Among the three, the Keough pool was the most concentrated and tightly organized, with a Herfindahl index (HHI, based on the size of farms in the pool) of 1,583. The Watterson pool had a Herfindahl index of 1,163, and the Cashbaugh, 410.[20]

Within the Keough pool, there was some defection, with seventeen of the twenty-three members selling in 1926 and 1927, but these were very small farmers (fourteen of them had ten acres each). The core of the pool—led by the largest landowner, Karl Keough, with 4,482 acres (60 percent) of the 7,862 acres on the Owens River Canal and by far the most water of any pool member—and five other farmers held out until 1931 for higher prices. For example, pool member G. L. Wallace offered his lands in 1926 for $417 per acre, while the city countered with $254 per acre. Final agreement was not reached with him until 1931 (when farm properties elsewhere in the country were falling because of the Great Depression) at $466 per acre.[21]

In 1926, the Keough pool demanded $2.1 million for its properties on the Owens River Canal. The Special Owens Valley Purchasing Committee, representing the LADWP, countered first with $1.025 million, then with $ 1.25 million, and finally with $1.6 million, but these offers were rejected.[22] By July 1929, the board was eager to buy the holdout farms, especially those on the Owens River Canal that had a lot of water, in order to secure supplies for the aqueduct. A. J. Ford, assistant right-of-way and land agent for the LADWP, and W. R. McCarthy, a representative of the holdout farmers, sent a letter to Edward D. Goodenough, a Ventura County supervisor, asking him to serve along with McCarthy and Ford as the third member of a price review committee.[23] His role was to help break the negotiation deadlock.[24]

Goodenough was to analyze the characteristics of the farms and to come up with a suggested price of his own. He reexamined the information assembled for each farm regarding location, size, alfalfa, orchards, pasture, livestock, brush, improvements, and water supplies (ditch shares, flow from creeks).[25] He also compared the prices asked by the farmers, the adjusted prices provided by Mc-Carthy, and those offered by Ford, before submitting his suggestions.

Table 4.1 lists four prices: those initially demanded by each farmer, Ford's offer prices for the LADWP, McCarthy's second offer on behalf of the farmers, and Goodenough's price offers as arbitrator. For key remaining properties, the prices reveal the gaps that remained even as late as 1929. As shown, Goodenough tended to suggest prices closer to those of McCarthy than of Ford. Considering the twenty-nine entries where the data are complete, Goodenough's prices were 79 percent of the initial asked price, whereas Ford's were 58 percent, and McCarthy's were 86 percent. The prices were not binding on the city, but were recommended. The last two columns of the table show the final sale prices and year the transactions were completed. The data indicate that the prices suggested by Goodenough were the final ones agreed to in most cases, but that agreement was not reached for up to two more years.[26] In the statistical analysis below, it is possible to isolate the characteristics of the farms, including Keough pool membership, that led to higher prices per acre.

The Cashbaugh and Watterson pools on Bishop Creek were less tightly organized, as the Herfindahl measures suggest, and they suffered from early sales by their largest landowners. Within the Watterson pool, the leaders and biggest landowners, Wilfred and Mark Watterson, with 1,216 acres across three separate plots, agreed to sell to the LADWP in 1926 at a slight premium over the agency's offer. They were quickly followed by all but three of the twenty pool members. The others sold in 1927.

The LADWP paid $1,389,364 to buy out the properties in the Cashbaugh pool between 1924 and 1927. Twenty of the forty-three members of the pool sold their farms by 1926. The largest landowner, William Cashbaugh, released his farm of 596 acres in 1927 to the city for $174,680, receiving a 21 percent premium over the LADWP's initial offer of $145,180.[27]

One factor behind the 1927 collapse of these two pools was the March 1927 board announcement that it would no longer buy properties in Owens Valley after May 1, 1927, and the failure of the Watterson brothers' bank.[28] Although

TABLE 4.1

Comparison of farm valuation differences, Owens Valley properties,
1929, and final sale prices (dollars)

Farmer	Initial price asked by owner	Offer price from city (A. J. Ford)	Price offered by farmers' arbitrator (McCarthy)	Final arbitration price (Goodenough)	Final sale price	Year of sale
Baker, F., E. P.[b]	108,000	75,330	104,944	98,624	98,624	1931
Craig, F. V.[a]	n.a.	6,300	6,696	6,696	n.a.	n.a.
Crosby, J. B.[c]	25,000	19,850	46,034	33,818	21,000	1929
Crow, W.H.B.[d]	18,000	13,395	22,834	18,998	n.a.	n.a.
Davies, A. M.[c]	n.a.	10,325	22,334	17,215	n.a.	n.a.
Detrick, W. E.[d]	19,200	13,680	17,164	16,515	n.a.	n.a.
Dumke, G. C.[c]	45,000	32,344	64,282	49,513	49,513	1931
English, F. H.[c]	29,600	22,235	36,768	30,551	30,551	1931
Ferris, W. H.[c]	75,000	38,075	69,692	69,504	n.a.	n.a.
Gifford, C. N.[c]	10,000	8,370	15,725	12,445	n.a.	n.a.
Harper, Jas.[c]	3,000	1,150	1,953	1,488	n.a.	n.a.
Hoskins, W. J.[c]	n.a.	16,190	25,878	21,070	n.a.	n.a.
Keough, K. P.[a]	165,000	124,793	190,651	179,401	179,401	1931
Math, Joe[a]	7,000	3,548	5,731	5,295	5,295	1931
Matlack, May[a]	n.a.	7,628	9,431	9,431	n.a.	n.a.
McIntyre, D.[a]	22,500	17,830	22,636	20,372	n.a.	n.a.
Morehouse, W.[a]	n.a.	3,750	6,091	5,603	n.a.	n.a.
Paget, R. S.[a]	59,000	39,820	50,319	50,748	n.a.	n.a.
Powers, A. E.[a]	35,000	24,530	38,261	34,435	n.a.	n.a.
Ripley, W.[a]	10,000	7,630	9,992	9,154	9,154	1931
Robb, G.[a]	9,660	8,550	9,667	9,108	9,108	1931
Rossi, J. W.[a]	25,000	19,767	19,837	20,460	n.a.	n.a.
Rowan, W. L.[a]	22,945	19,350	21,628	20,495	n.a.	n.a.
Shaw, Bob[a]	n.a.	4,170	6,172	5,710	n.a.	n.a.
Shaw, Maria[a]	n.a.	4,420	6,185	5,937	n.a.	n.a.
Smith, A. J.[c]	5,000	2,790	6,000	5,078	5,078	1931
Sommerville, M.[b]	24,000	21,436	24,111	23,778	n.a.	n.a.
Stofflet, J. H.[a]	30,000	18,170	22,341	20,480	20,480	1931
Tichenor, Jesse[a]	20,000	16,693	19,386	18,375	17,701	1931
Trickey, W. A.[d]	21,567	21,410	28,484	24,556	n.a.	n.a.
Veil, Joe[b]	n.a.	9,375	22,996	19,884	n.a.	n.a.
Wallace, G. A.[a]	28,0001	25,980	31,453	28,308	28,509	1931
Wallace, G. L.[a]	52,500	49,515	62,299	56,070	55,964	1931
Watson, C. P.[b]	57,500	26,295	42,356	39,476	n.a.	n.a.
Watterson Bros.[a]	131,500	83,390	143,593	136,414	n.a.	n.a.
Webb, Thos.[b]	21,000	9,960	11,930	11,349	11,349	1931
White, Olive[a]	5,000	3,875	4,833	4,528	4,528	1931

SOURCES: Property List, 1929, Goodenough Papers, Eastern California Museum, Independence, Calif.; Final sales prices and year are from "Tabulation Showing Status of Ranch Land Purchases Made by the City of Los Angeles in the Owens River Drainage Area from 1916 to April 1934," prepared in Right of Way and Land Division by Clarence S. Hill, Right of Way and Land Agency, compiled by E. H. Porter, April 16, 1934, Tape GX0004, LADWP Archives.

NOTES: n.a. = not available.

[a] Owens River Canal.

[b] Big Pine area.

[c] Laws-Chalfant area.

[d] West Bishop area.

the board later resumed land purchases in early 1929, this action raised uncertainty about whether further acquisitions might end or at least be delayed. The collapse of the bank, at a time when rural banks were failing all over the western United States, also weakened the resolve of the pool members. The Wattersons, who were important leaders in the negotiations with the LADWP, were charged with embezzlement, tried in November 1927, found guilty, and sentenced to ten years in San Quentin prison.[29]

Nonpool Ditch Farms

In order to block the formation of the Owens Valley Irrigation District, the LADWP in 1923 and 1924 raised its offer prices for farms on two key ditches, McNally and Big Pine, and quickly purchased them at a premium. This action broke the geographic continuity and membership cohesiveness of the irrigation district, and by the end of March 1925 it was effectively sidelined. It will be possible in the statistical analysis to determine how these farmers did relative to their neighbors who were or were not in pools. Conceptually at least, these farmers would not have sold early unless they could earn at least what they expected to by holding out.

Unorganized Competitive Nonditch Farms

Nonpool farmers, who were not on key ditches, competed to sell their properties to the board, often writing to it with their offers. In 1925, Board Counsel W. B. Mathews commented on the "insistent demand" by some property owners for the city to buy their lands.[30] This competition revealed reservation values, and there is no evidence that these farmers were involved in any of the bargaining conflicts in Owens Valley. Indeed, in these negotiations, the board reported that "the prices paid, with few exceptions, have been entirely satisfactory to the seller."[31] For this reason, the analysis of bargaining conflicts focuses on the LADWP and farmers who were in the sellers' pools.

Table 4.2 provides comparative information on the pool farms, those ditch farms purchased preemptively to block formation of the Owens Valley Irrigation District, and nonditch farms.

As shown in the table, pool farms tended to be more homogeneous with respect to size, fraction cultivated, water endowments, and riparian position than either the nonpool ditch farms or the more scattered nonditch farms. In terms of mean price, pool members received more per acre of land than their counterparts either on ditches or elsewhere in Owens Valley.

TABLE 4.2
*Characteristics of pool member farms, nonpool ditch farms,
and nonpool, nonditch farms in Owens Valley*

Farm type (number of farms)	Farm size (acres)	Fraction cultivated	Water acre-fee/acre	Riparian (0,1)	Price/acre (dollars)	Year of purchase
Pool farms (86)						
Minimum	10.00	0	0	0	17.66	1925
Maximum	808.00	1.00	11.80	1	954.90	1931
Mean	118.14	0.173	4.64	.41	294.60	1927
Std. deviation	133.84	0.210	2.39	.49	163.74	1.29
Coefficient of variation	1.13	1.21	0.52	1.20	0.56	—
Nonpool ditch farms (281)						
Minimum	10.00	0	0	0	20.00	1922
Maximum	3,502.41	1.00	16.50	1	766.20	1931
Mean	121.94	0.224	5.35	.36	263.32	1926
Std. deviation	274.69	0.301	2.74	.48	152.81	1.73
Coefficient of variation	2.25	1.34	0.51	1.33	0.58	—
Nonpool, nonditch farms (228)						
Minimum	10.00	0	0	0	3.00	1917
Maximum	2,364.40	1.00	7.00	1	620.99	1932
Mean	207.03	0.093	1.19	.31	81.62	1927
Std. deviation	287.46	0.197	1.93	.46	92.68	1.97
Coefficient of variation	1.39	2.12	1.62	1.48	1.14	—

SOURCE: Data set of farms ten acres or larger drawn from "Tabulation Showing Status of Ranch Land Purchases Made by the City of Los Angeles in the Owens River Drainage Area from 1916 to April 1934," prepared in Right of Way and Land Division by Clarence S. Hill, Right of Way and Land Agency, compiled by E. H. Porter, April 16, 1934, Tape GX0004, LADWP Archives.

PRICE DISPUTES

As pool farmers negotiated with the LADWP, there were two sources of price disputes. One was the value of any particular property when farms were heterogeneous. Differences between a bid and an asking price could sometimes be very large. One farmer rejected a bid of $11,200 for his fifty-acre farm, claiming that the agency's assessment undervalued the water and improvements on his property. Using prices paid by the LADWP for neighboring properties with and without water, he estimated the added value of water, incorporated it into his calculation, and countered with an asking price of $18,339. He then held out for two years, selling the farm to the city for $19,000.

The second issue was the value of the water on a farm: whether it should be valued as an input to agricultural production in Owens Valley, as desired by the board, or as an input to land value increases in Los Angeles, as desired by farm-ers. For example, before the board, one landowner claimed that she priced ac-cording to "the comparative value of what that water is worth to you . . . . because we know you want water and not the land . . . that is what you want and all you want."[32] Given the observed land value increases in Los Angeles after the arrival of Owens Valley water, relative to the value of land if sold for agricultural use in Owens Valley, there were considerable gaps between offered and demanded prices for the farms with the most water.

A great deal of water was concentrated in certain parts of Owens Valley while other parts were comparatively dry. With limited arable land throughout the valley, not all of the water in well-endowed areas could be translated into ad-ditional cultivated acreage and significant agricultural production. As a result, water-intensive flood irrigation was common on the limited land available, and early observers commented on the profligate use of water by Owens Valley farm-ers, which left some lands water-logged.[33]

A regression of total cultivated acreage per farm on water acre-feet illustrates the situation. The estimated coefficients indicate that an additional water acre-foot available to a farmer allowed for an increase in cultivated land of only .035 acres in Owens Valley.[34] This is less than in contemporary agriculture elsewhere in California where an additional acre-foot of water adds from .17 to .33 acres of cultivation.[35] As further illustrated below, the marginal agricultural value of extra water on many Owens Valley farms was low. For this reason, disputes over the valuation of water were particularly contentious for farms with the most wa-ter because on those properties marginal values were closer to zero than on com-paratively drier farms. Indeed, farms in two of the pools had the most water per cultivated acre of land in the valley.

EMPIRICAL ANALYSIS

As noted above, the LADWP Archives has a data set of 869 farms purchased in the northern Owens Valley between 1916 and 1934. Excluding properties of ten acres or less as not being farms, but town lots, as well as dropping incomplete en-tries, leaves 595 observations. These properties are described in Table 4.3, which reports the mean values from the data set.

Of the 595 farms, 367 were on irrigation ditches; 228 were not on ditches, but scattered throughout Owens Valley. For all farms, the mean water acre-feet per cultivated acre of land was 28, although this figure ranged from 18 to 69 acre-feet per cultivated acre among farms on irrigation ditches and was 14 acre-feet

TABLE 4.3
Characteristics of Owens Valley farm property, mean values

Property type	Price/ acre (dollars)	Total purchase price (dollars)	Year of purchase	Size (acres)	Price/ water acre-foot (dollars)	Total water acre-feet
All properties	198	23,425	1926	154	178	448
Farms not on ditch	82	19,890	1927	207	473	261
Keough pool	443	27,647	1928	79	77	366
Cashbaugh pool	242	32,156	1927	126	69	544
Watterson pool	237	33,983	1926	147	75	584
Nonpool on ditches	263	23,861	1926	122	112	581

	Water acre-feet/acre	Total cultivated acreage	Cultivated land (percent)	Water acre-feet/ cultivated acre	Riparian rights (percent)	Number of farms
All properties	4	17	17	28	35	595
Farms not on ditch	1	19	9	14	31	228
Keough pool	6	16	20	69	96	23
Cashbaugh pool	4	15	14	33	19	43
Watterson pool	4	27	21	18	25	20
Nonpool on ditches	5	14	22	30	36	281

SOURCE: "Tabulation Showing Status of Ranch Land Purchases Made by the City of Los Angeles in the Owens River Drainage Area from 1916 to April 1934," prepared in Right of Way and Land Division by Clarence S. Hill, Right of Way and Land Agency, compiled by E. H. Porter, April 16, 1934, Tape GX0004, LADWP Archives.

per cultivated acre for farms located on drier areas not on ditches. These figures represent the stock of water available annually with each farm, and because production generally involved only 180 days or less, actual irrigation might have required about half of the quantities listed in the table. Even so, the amounts represent considerable quantities of water available for each acre in cultivation for many of the farms in Owens Valley.

Mean Values for Owens Valley Farms, by Category

Farms on ditches sold for higher prices per acre and greater total prices than those that were not on a ditch (see Table 4.3). The former had higher percentages of cultivated land and more water per acre of land, and their owners were more likely to be in a sellers' pool. Those farmers who were in the Keough pool commanded the highest price per acre of land; they sold the latest (held out the longest); and they had the most water per acre to offer Los Angeles. On average, members of the other two pools, Cashbaugh and Watterson, also did better than did nonditch farmers in terms of price per acre and total purchase price. Even nonpool members who were on ditches earned more in total and per acre of land than the nonditch farmers. These farmers benefited from the early actions of the LADWP to purchase their farms before joining a pool.

In contrast, nonditch properties sold for less in total and per acre of land. They typically had a smaller share of cultivated farmland and carried less water, and their owners were unorganized. Although they received less for their land, these farmers earned more per water acre-foot than farmers more favorably located on ditches. This outcome reflects the purchase of a bundled asset in the land market. At a minimum, the LADWP had to pay a price that equaled the agricultural value of a farm in order to secure it and its water from the owner. If not, all additional water on a farm translated directly into increased agricultural production; then farmers with less water were likely to receive more per unit of water than their counterparts who had larger water endowments.

This issue is examined in the econometric analysis below, but the mean values shown in Table 4.3 are suggestive that added water increased farm values at a declining rate. For nonditch, less productive farms, the average farm sale price was $19,890, and this translates to an implicit price of $473 per acre-foot of water (farm sale price per water acre-foot). The total farm sale price is somewhat less than the mean 1925 census farm value for the four comparable Great Basin counties (Lassen, California; Churchill, Douglas, and Lyon, Nevada) of $21,167, but these nonditch farms were the least productive properties in Owens Valley. Their mean sale value of nearly $20,000 corresponded to six years of gross farm receipts for Inyo County farms during a time of agricultural depression.[36] It is no wonder then that nonditch farmers sold quickly whenever they had an offer, without discord.

The mean sales value for pool farms on ditches was higher than for nonpool farms, and all ditch farmers received more in total than their nonditch colleagues. The sale prices for both pool and nonpool ditch farms were considerably higher than the 1925 mean census farm value for the four Great Basin counties identified above. The farmers on these properties also received per acre land prices that were at least three times those obtained by those not on ditches. Importantly, however, ditch farmers who had the most water received less per unit of water in their negotiations with the LADWP.

ESTIMATION FRAMEWORK

Regression analysis allows for more precise examination of the factors that determined the year of sale and the price paid per acre of land and per acre-foot of water. Based on discussion of the bargaining environment, the following relationships are estimates, first for year of sale:

(1) Year of sale = g (farm size, farm size^2, cultivated acreage, cultivated acreage^2, water/acre, water/acre^2, riparian water rights, pool membership, nonpool ditch farms, change in Los Angeles's population from the previous year, and current aqueduct flow per capita).

The data set described below includes only the year in which each property was sold to the Water Board. In does not, in general, include when initial offers were made or the amounts sought by both sides. Accordingly, the observed year of sale reflects both demand factors, indicating the interests of the LADWP, and supply conditions, indicating the interests and efforts of farmers. The agency was charged with providing water to the city. Greater changes in Los Angeles's population from the previous year and declines in the current flow of aqueduct water per capita would encourage the board to return to Owens Valley to buy more properties to maintain water supplies.[37] Farms with more water per acre and riparian water rights would be sought earlier to meet this demand. Further, all else equal, larger farms would have more water, and those with more cultivated acreage may have signaled access to water not captured in the other water variables, suggesting that these two variables could lead to earlier years of sale. These factors also may have made farms more commercially viable, allowing farmers to hold out longer. Accordingly, the overall effects of the variables are ambiguous. The squared terms are included to address nonlinearity. On the other hand, farmer pool membership should delay sale according to the strength of the pool. The nonpool ditch farmers who were early defectors from the irrigation district as described above should also have earlier years of sale, relative to the baseline of competitive nonpool, nonditch farms.

(2) $P = f$ (farm size, farm size^2, cultivated acreage, cultivated acreage^2, water/acre, water/acre^2, riparian water rights, pool membership, nonpool ditch farms, cumulative percentage of total water purchased).

The recorded per acre price paid for farmland in Owens Valley also reflects both demand and supply factors. Those variables that measure agricultural productivity—farm size (economies of scale), cultivated acreage (other water, inherent fertility, topography), measured water per acre, and riparian rights—would raise prices by increasing the reservation values of farmers. To reduce overall transaction costs, the board also would be willing to pay more for farms with more water and that were larger. The squared terms, however, are likely

TABLE 4.4
Descriptive statistics for Owens Valley farms

Variable (595 observations)	Mean	Standard deviation	Minimum	Maximum
Land price/acre (dollars)	198	163	3.00	955
Year of purchase	1926	1.87	1917	1932
Farm size (acres)	154	267	10	3,502
Cultivated acreage	17	40	0	421
Water acre-feet/acre	3.7	3	0	16.5
Riparian rights (Y/N)	0.35	0.48	0	1
Keough pool (Y/N)	0.04	0.19	0	1
Cashbaugh pool (Y/N)	0.07	0.26	0	1
Watterson pool (Y/N)	0.03	0.18	0	1
Other ditch (nonpool) (Y/N)	0.47	0.50	0	1
Lagged Los Angeles population change (in 1,000s), 1916–34	123	75	23	283
Annual aqueduct flow/Los Angeles population	0.00016	0.00003	0.00012	0.00032

to be negative, reflecting a falloff in the marginal effects of these two variables. The pool member variables capture the relative bargaining strengths of the three pools, also raising price, and the nonpool ditch farms variable captures the effect of board purchases to encourage defection from the irrigation district, relative to the competitive fringe, nonpool, nonditch farms. The cumulative percentage of total water purchased as of each sale also raises farm prices because it reflects a growing supply constraint faced by the LADWP in its negotiations.

Although the observed market trades were for land, it is possible to calculate an implicit price of water and estimate the determinants of water prices using the same variables outlined in (2) above. This estimation will illustrate whether bargaining power in the land market translated into higher prices for water.

Regression Analysis of the Determinants of Year of Sale and Land and Water Values in Owens Valley

We first examine the land market where the actual negotiations took place.

Statistical analysis of the land market: year of sale and price per acre. Table 4.4 provides descriptive statistics for an econometric analysis of the bargaining over Owens Valley lands.

In the regression analysis, the year of sale is regressed against water per acre (predicted effect − earlier sale), water/acre^2 (+), farm size (−/+), farm size^2(+/−), cultivated acreage (−/+), cultivated acreage^2 (+/−), riparian water rights (−), pool membership (+), nonpool ditch farms (−), lagged change in Los

TABLE 4.5
The land market in Owens Valley: year of purchase

Variable	Coefficient	Std. error
Constant	1927.41***	0.42
Total farm acres$_t$	$-2.0E^{-03}$***	5.4E^{-04}
Total farm acres^2	5.6E^{-07}**	2.3E^{-07}
Total cultivated acreage$_t$	0.02***	0.004
Total cultivated acreage^2	$-2.2E^{-05}$*	1.4E^{-05}
Water acre-feet/acre$_t$	-0.26***	0.06
Water acre-feet/acre$_t$^{2}	0.01***	0.006
Riparian rights$_t$	0.14	0.15
Member of Keough pool$_t$	1.08***	0.38
Member of Cashbaugh pool$_t$	-0.20	0.29
Member of Watterson pool$_t$	-0.60	0.39
Farms on ditches but not in pool$_t$	-0.52***	0.20
Aqueduct flow/L.A. population$_t$	8638.62***	2123.50
Change in Los Angeles population$_{t-1}$	-0.01***	0.002

NOTES: 595 observations, $R^2 = .28$, $F(13, 581) = 18.74$.

*** significant at the 1 percent level or better.

** significant at the 5 percent level.

* significant at the 10 percent level.

Angeles's population ($-$), and current aqueduct flow per capita ($-$). For ease of interpretation, OLS (ordinary least squares) is used.[38] The results are provided in Table 4.5.

The findings from this exercise provide insight into the factors that determined when properties were sold in Owens Valley. As shown in the table, among the sellers' pools, members of the Keough pool on average delayed sale a year longer than the baseline farmers not on ditches. For members of the Watterson and Cashbaugh pools, however, sale time is not statistically distinguishable from the baseline group. But farmers who were on ditches but not in pools, and who had their properties purchased early by the LADWP to halt the formation of the irrigation district, sold about half a year earlier than the baseline. Farms with more water were purchased earlier, reflecting the agency's desire to secure properties that brought more water to the aqueduct. An additional acre-foot of water per acre of land speeded sale by .26 year, or about three months. This effect, however, was mitigated for a few farms with very large water holdings. The estimated coefficients on the water per acre and water per acre squared terms suggest that at 13 acre-feet per acre, more water would no longer speed sale. Of the 595 farms, four had endowments of that size.

A past increase in Los Angeles's population of 100,000 people prompted earlier sales by a year. In contrast, a rise in current aqueduct flow per capita, indicating

TABLE 4.6
The land market in Owens Valley: price per acre

Variable	Coefficient	Std. error
Constant	-49.33***	15.88
Cumulative water$_t$	144.55***	15.89
Total farm acres$_t$	-0.15***	0.04
Total farm acres^2	$3.7E^{-05}$**	$1.5E^{-05}$
Total cultivated acreage$_t$	0.69***	0.25
Total cultivated acreage^2	$-1.7E^{-03}$**	$8.7E^{-04}$
Water acre-feet/acre$_t$	37.52***	4.13
Water acre-feet/acre$_t$^{2}	-1.12***	0.37
Riparian rights$_t$	-1.45	9.56
Member of Keough pool$_t$	213.22***	24.69
Member of Cashbaugh pool$_t$	52.34***	18.66
Member of Watterson pool$_t$	81.13***	24.92
Farms on ditches but not in pool$_t$	68.64***	12.60

NOTES: 595 observations, $R^2 = .62$, $F(12, 582) = 78.48$.
*** significant at the 1 percent level or better.
** significant at the 5 percent level.
* significant at the 10 percent level.

a greater water supply, delayed sale, although more information is required to interpret the coefficient. The variable is measured in cubic feet per second per Los Angeles population and is very small, with a mean of .00016. The change between the largest and smallest flow per capita over the period of analysis is .0002, and multiplying that figure times the coefficient gives 1.76. This suggests that Los Angeles bought farms about two years later when the aqueduct flow was at its highest level relative to its lowest level, all else equal.

Turning now to analysis of the price per acre of land, price is regressed against water per acre (predicted effect +), water/acre^2 (−), farm size (+), farm size^2 (−), cultivated acreage (+), cultivated acreage^2 (−), the existence of riparian water rights (+), membership in one of the three sellers' pools (+, with the effect varying according to strength of the pool), whether a property was a nonpool ditch farm (+), and the cumulative percentage of total Owens Valley water purchased each year (+). Ordinary least squares is used in the estimation.[39]

Table 4.6 provides the estimated results for price per acre.

Again, the findings are informative regarding the determinants of the prices paid for land in Owens Valley. As shown in the table, the estimated coefficient for the supply constraint variable facing the LADWP, the cumulative percentage of total water purchased as of each sale, is approximately 145. This suggests that a 1 percentage point increase in the cumulative amount of water purchased relative to the total available raised the price of land by $1.45 per acre. Among the agricultural productivity variables, water endowments mattered the most, with an additional acre-foot of water per acre of land adding over $37 per acre to the sale price. This contribution, however, grew at a declining rate. The falloff in the value of the marginal product of additional water per acre varied across the sample, with the farms at the center of the most contested negotiations having the largest negative effects. The size of the estimated coefficients on the water per acre and water per acre squared terms implies that the value of the marginal product of water per acre was zero at 16.8 acre-feet per acre. In the data set analyzed here, one farm with the largest water holdings had 16.5 acre-feet per acre; three had 13 acre-feet per acre or more; and thirty farmers had at least 9 acre-feet per acre. Hence those farmers with the most water had the greatest reason to resist efforts by the LADWP to purchase water-bearing properties according to their agricultural productivity values. These were the farmers who colluded to secure higher per acre prices.

Sellers' pools exhibited market power in the land market, with members of the Keough pool earning about $213 more per acre than the 228 nonditch property owners and $145 more per acre than farmers who were on ditches but had defected from collusive efforts. Members of the Watterson and Cashbaugh pools earned approximately $81 and $52 more per acre respectively than the competitive baseline farmers. To keep farmers out of the irrigation district, Los Angeles paid an additional $68 per acre for the nonpool ditch farms, an amount more than their owners would have received in the Cashbaugh pool, but less than in the Watterson pool.[40]

There is additional information regarding the relative bargaining strength of the sellers' pools for 135 farms in Owens Valley on farmers' asking prices, the offer prices from Los Angeles, the final prices paid, and appraisal values. Although the data are not consistently provided for all the observations across the groups, there is enough information for further assessment of the relative effectiveness of pool members in their negotiations with the LADWP.

The data in Table 4.7 are suggestive of the price adjustment process by the Water Board in bargaining with pool members. For the less effective Cashbaugh pool, the price offered by the board was virtually equal to the adjusted appraisal values (4.1 times appraisal according to the pricing rule), whereas for the more intransigent Keough pool members, the LADWP raised its offer by 45 percent on average from the appraisal value in an effort to secure the properties.

Table 4.8 reports the results of a regression of the ratio of the asking to final purchase price against a constant and a number of control variables, including pool membership for 135 farms. The results indicate that Keough pool members

TABLE 4.7
Analysis of bid prices and appraisal values by owner group

	Cashbaugh pool properties (N = 35)	Keough pool properties (N = 8)
Mean ratio of L.A. bid prices to L.A. appraisal values	.99	1.45

SOURCE: "McNally Ditch Stockholders," Appraisal files, Tape GX0008; "Sale of Lands," 1920–27 file, Keough Pool, Owens River Canal; Miscellaneous File, Appraisal and Ask Values; 1927 Appraisal Values, Tape GX0004; "Fish Slough" file, Cashbaugh Pool, Tape GX0001; McNally Ditch, Tape GX0002; all in LADWP Archives.

TABLE 4.8
Analysis of asking-to-final-purchase-price ratio by owner group

Variable	Coefficient	Std. error
Constant	−86.37	28.56
Total farm acres	−0.0002	0.0002
Year of purchase	0.05	0.01
Keough pool	−0.33	0.12
Cashbaugh pool	−0.21	0.11

NOTES: 135 observations, $R^2 = .07$, $F_{(4, 130)} = 3.66$.

Dependent variable = ratio of asking price to final purchase price.

received more than they initially asked for. They were able to do so by successfully holding out for an even higher price whenever the board rejected their initial demands. The mean asking-to-purchase-price ratio for the entire group of 135 farmers was 1.19. Using this ratio, the coefficient results suggest that the Keough pool's mean asking-to-final-price ratio was .86, while the Cashbaugh's mean ratio was .98.

All told, the analyses of the year of purchase, the purchase price per acre, and price negotiations provide important insights into the Owens Valley controversy. The LADWP and the farmers battled over the value of the water-bearing lands. The agency was forced to pay premium prices for some ditch properties in order to prevent the formation of the Owens Valley Irrigation District as a single negotiating unit. The data in the tables show that the agency was right to be concerned. Pool farmers did better than their unorganized, nonpool colleagues, and the Keough pool did especially well in the land market. A single unit might have done even better. Conflicts over price, however, under bilateral monopoly conditions helped to give the Owens Valley water transfer its contentious history.

TABLE 4.9
The water market: price per acre-foot

Variable	Coefficient	Std. error
Constant	−1457.73***	288.65
Cumulative water$_t$	−67.16	178.50
Total farm acres$_t$	−1.05**	0.41
Total farm acres^2	3.1E^{-04}*	1.7E^{-04}
Total cultivated acreage$_t$	2.97	2.65
Total cultivated acreage^2	−8.4E^{-03}	8.9E^{-03}
Water acre-feet/acre$_t$	−358.53***	55.37
Water acre-feet/acre$_t$^{2}	22.69***	4.36
Riparian rights$_t$	26.33	107.58
Member of Keough pool$_t$	−14.65	265.67
Member of Cashbaugh pool$_t$	−194.29	192.91
Member of Watterson pool$_t$	−287.34	248.72
Farms on ditches but not in pool$_t$	−108.84	138.27

NOTES: 443 observations, R^2 = .09, $F(12, 430)$ = 4.86.
*** significant at the 1 percent level or better.
** significant at the 5 per cent level.
* significant at the 10 percent level.

Statistical analysis of the water market. We now turn to an analysis of the hypothetical water market. It is hypothetical because we do not actually observe these transactions. Table 4.9 reports an OLS estimation of the implicit price per acre-foot of water. The price of an acre-foot of water is obtained by dividing the total sale price for a farm by the amount of water associated with it. Because not all farms had water endowments, the sample sizes in the land and water market analyses are not the same.

Overall, the model performs poorly in explaining the implicit price of water, suggesting that there was only a weak relationship between the variables that affected land prices and the prices received per acre-foot of water. After all, the LADWP was using the land market to gain water so that the relationship would be imperfect. Even so, the results help explain the lingering notion of "theft" in Owens Valley. There is no statistically significant effect of pool membership on water prices. While pool members earned more per acre of land, their collusive ability was not enough to translate into higher water prices. Importantly, the water per acre variable has a significant and negative coefficient of −358.53. The more water a farm conveyed with its purchase, the lower the water price per acre-foot. Although this negative effect was mitigated as the amount of water on a farm increased, the size of the estimated coefficients on the water per acre and water per acre squared terms indicates that the effect did not become zero

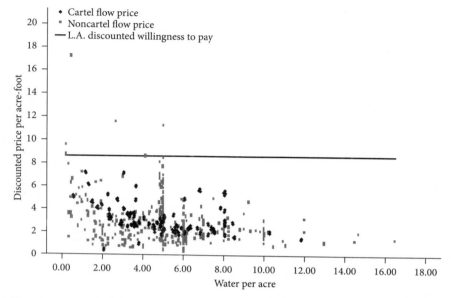

Figure 4.3 Flow Prices Paid to Owens Valley Farmers Relative to What Los Angeles Paid for Colorado River Water.

SOURCE: "Tabulation Showing Status of Ranch Land Purchases Made by the City of Los Angeles in the Owens River Drainage Area from 1916 to April 1934," Prepared in Right of Way and Land Division by Clarence S. Hill, Right of Way and Land Agency, Compiled by E. H. Porter, April 16, 1934, Tape GX0004, LADWP Archives.

(and then turn positive) until water endowments reached 7.9 acre-feet. Only 55 of the 443 farms with water in Owens Valley had this amount of water or more. For most farmers, then, the more water they had, the lower the water price they received.

There is another indication of the relative market power of the two parties in the negotiations over water-bearing land. It is possible to compare the implicit prices paid per water acre-foot with the price that the LADWP might have been willing to pay. In 1931, voters in the Metropolitan Water District, which included Los Angeles, approved bonding for $220 million for construction of the Colorado River Aqueduct to bring 1.1 million acre-feet to the city annually. This translates approximately to $220 per acre-foot for water from the Colorado River, or $9.50 per acre-foot for an annual flow.[41]

Similarly, converting all implicit water prices for each Owens Valley farm into prices for an annual flow of water and plotting them in Figure 4.3 illustrates the position of the farmers relative to this baseline. As shown, farmers generally received well below the maximum amount the board might have been willing to pay, regardless of whether a farmer was part of a sellers' pool.

Moreover, consider the total expenditures made by the LADWP relative to what it might have been willing to pay. The total outlay for Owens Valley farms by the LADWP between 1916 and 1934 in the data set used here was $13,937,934, where 266,428 acre-feet of water were secured with an average price per acre-foot of $52.31.[42] Part of the $24.6 million capital cost of the Los Angeles Aqueduct should be added to this figure because it was necessary to move the water. The LADWP purchased farms and their water after 1916 in order to expand the aqueduct flow close to capacity, from 290 to 460 cubic feet per second (cfs).[43] The increase was 170 cfs, or 37 percent of aqueduct capacity. Adding 37 percent of the capital cost, or $9,102,000, to the land expenditure figure results in a total outlay of $23,076,934 for water and relevant capital costs.

If the LADWP had paid $220 per acre-foot for Owens Valley water, as it did for Colorado River water, the total sales expenditures would have been $58,614,375, or about two and a half times the actual outlay. For even the farms that the agency bought preemptively in 1923 and 1924 to block the Owens Valley Irrigation District, the price paid per acre-foot was $68.50.[44] It seems then that the LADWP paid less for Owens Valley water than it had to pay for Colorado River water and hence what it might have been prepared to pay. This is only a suggestive exercise, but the outcome reflects the relative bargaining power of the agency, and it strengthens the sense that water was stolen. Farmers did better by selling their farms than if they had stayed in agriculture, but most of the surplus from transferring water from agriculture to urban use went to Los Angeles property owners.

THIRD-PARTY EFFECTS: TOWN LOT VALUES

A major bargaining issue facing the LADWP and Owens Valley residents involved what to do about the town properties in the five towns of Bishop, Big Pine, Lone Pine, Independence, and Laws. As noted in Chapter Three, property owners complained that the commercial value of their town lots was reduced by the city's acquisition of farmland and gradual reduction of irrigated agriculture. This is an example of the complicated and contentious third-party effects commonly encountered in western water transactions.

Although the magnitude of the effects was disputed by both parties, enactment of the 1925 legislation, subsequent damage and reparations demands by town property owners, and a 1929 California Supreme Court ruling validating the 1925 law forced the LADWP to begin negotiations to acquire the town properties.[45] But just as with farmland, town lot owners proved to be adept negotiators by forming sellers' pools (the Johnson-Riley or "Main Street pool" in Bishop, for example), enlisting added support from the state legislature, and forcing the LADWP ultimately to offer very favorable terms for the properties.

Chapter 109 of the Statutes and Amendments to the Codes of California for 1925 "imposed liability upon any municipal corporation or upon any person, firm or corporation engaged in supplying water to any municipal corporation that enters any watershed or any lands, streams or waters in the watershed, for the purpose of acquiring or increasing a water supply for such purpose, and taking, diverting or distributing water for use by a municipal corporation" for damages to adjacent properties.[46] Armed with this legislation, Keough, Watterson, and Cashbaugh pool negotiators used the issue of the town lots to bring additional pressure on the LADWP to agree to their price demands. The alleged deterioration of property values in the towns due to the transfer of water and decline of irrigated agriculture served to bolster their argument that using local agricultural values to determine pricing, as insisted on by the LADWP, was inappropriate. Pool members wanted the water values in Los Angeles to be considered in setting the sale prices for their properties.[47]

As noted in Chapter Three, in 1925 the Owens Valley Reparations Committee demanded either that the Board of Water and Power Commissioners pay reparations for the loss in property values or purchase them for $12 million.[48] There was no obvious basis for the figure, and it was rejected by the board. In 1927 the reparation demands made by the Owens Valley Reparations Committee and the Big Pine Reparations Association were reduced, but these also were denied by the board.[49] No action was taken by the board until the courts ruled on the constitutionality of the 1925 state law requiring compensation for financial damages to businesses and other urban properties due to the transfer of water from a region by a municipality. Court opinions in 1929 and 1930 upheld the validity of the law and the proposed purchase of town lots in Owens Valley by the Los Angeles Water Board.[50]

On December 31, 1929, the Board of Water and Power Commissioners passed the following resolution:

WHEREAS, ... in pursuance of a policy of protecting and increasing the water supply of the City of Los Angeles, after a survey of all properties within the Towns of Bishop, Big Pine, Laws, Independence, and Lone Pine, in the

Owens River Valley, State of California, and after conferences between representatives of said Towns and of said Board, adopted a resolution providing, among other things, as follows:

Said Board offers to purchase from the owners willing and desiring to sell only the water and water rights in, on, under, appurtenant to, part of, or used in connection with, their lands, which is in excess of such surface water as may be needed for domestic purposes on said land, together with the right to extract, take and exhaust said waters at any point on or outside of said lands where said City or Board may have the right of access thereto, and to convey the same to any point in or outside of the Owens River watershed.[51]

To determine prices, a Committee of Ten was set up consisting of five representatives (one from each town) and five members of a Special Owens Valley Committee of the Board. But the property value appraisals prepared by the agency's land agents were rejected by the town representatives. Negotiators for the towns offered counterappraisals that raised proposed values in the towns of Laws and Independence by 45 to 50 percent, in Bishop by 120 percent, and in Big Pine by 60 percent. Members of the California legislature who owned properties in the towns, Senators Joe Riley and Dan Williams, threatened new legislative investigations of the city's purchasing practices unless the new appraisals were accepted.[52] Riley, Williams, and other members of the Main Street pool in Bishop demanded that the LADWP accept 1923 values for the properties, rather than 1929 values. Riley stated: "It is not my fault that the City of Los Angeles needs the water from this district in order that said City might be increased in value by millions of dollars, and it is no more than right that I should be compensated for the personal damage done ... it is not the time nor place for any City who needs water, to arbitrarily set a price on a man's property or simply guess as to what the price ought to be."[53]

By this time, Riley was quite confident that he would be able to extract a higher price from the city. The higher values he was holding out for were based on an 8 percent annual increase in town property values from those that were recorded in 1923. The 8 percent figure came from the observed growth rates in nine southern California towns that benefited from Owens Valley water. This was the rate of increase demanded by representatives of Bishop on the Committee of Ten. The agency's representatives countered that the growth in these southern California towns was due to their climate, oil, citrus, and access to shipping, and that trends in property values in thirteen other California towns more comparable to those in Owens Valley showed *declines* of 14 percent between 1922 and 1929. Another source of conflict for some town lot owners was the LADWP's insistence that any completed deal include a statement by the seller agreeing to drop "all legal suits or claims" for reparations against Los Angeles and would never object to its withdrawing "its *own* water from the Owens River watershed into the City of Los Angeles."[54] This requirement ultimately was dropped.

While these negotiations were under way in 1931, the California Senate passed a resolution prepared by the Senate Conservation Committee to investigate Los Angeles's purchase of land in Inyo and Mono Counties to determine the depreciation in town values caused by the city's actions:

> Whereas, The city of Los Angeles has diverted and appropriated certain waters of Inyo County and Mono County to the use and benefit of the city of Los Angeles. . . . and Whereas, It has been represented that the taking of said waters has resulted in the depreciation of the value of certain lands. . . . Whereas, It is advisable that a committee be appointed to study these maters and to report back its findings. . . . Resolved, further, That said committee is hereby authorized and empowered to do any and all things necessary to make a full and complete investigation.[55]

An investigating committee arrived in the valley and conducted hearings in 1931, allowing valley residents a forum to air their grievances. The findings were issued May 6, 1931, and were critical of the city:

> The city of Los Angeles buys the lands for the water only, and the water is taken into the aqueduct. The rich lands formerly irrigated by this water are dried up and allowed to return to desert conditions. . . . in a short time, lands that were rich and productive have returned to their former condition. . . . Damage to businesses, houses and property in the towns that were supported or were dependent upon the farm areas is very apparent. These towns are dwindling and gradually dying. Business is lost.[56]

The committee criticized the appraisal process used by the LADWP and its refusal to adequately value business property in the five towns and to purchase the equipment and fixtures used in those businesses. The committee called on the LADWP to fairly appraise the properties and to purchase all remaining water and lands in Owens Valley: "Your committee finds that the city of Los Angeles because of its wealth and power . . . as contrasted with that of the farmers and other property owners . . . who are, in the majority of cases people of small means and influence."[57]

By bringing in their colleagues from the legislature, Senators Riley and Williams were able to ratchet up the pressure on the LADWP to agree to terms. As we will see in Chapter Five, there is little evidence that Los Angeles's purchases of properties in Owens Valley had much negative impact on the town of Bishop, at least. California State Board of Equalization annual reports on the total value of all property within municipalities (real estate, personal, and monetary) support the claims of the LADWP. In 1920, prior to major purchases in the agricultural part of the valley near Bishop, the total value of property in Bishop was listed as $1,027,792. By 1929 it had risen to $1,355,666 (a gain of 32 percent), and even in 1930 with the onslaught of the Great Depression, town property in Bishop was estimated to be $1,228,709, a gain of 20 percent over 1920.[58] Accordingly, any negative third-party effects may have been quite small.

Adding to the legislative pressure on Los Angeles, Bishop residents sought court injunctions to halt groundwater pumping in the northern part of the valley by the city. The strategic use of these suits is indicated in a February 11, 1931, telegram from land agent E. A. Porter to the head of the Land Division, A. J. Ford. Porter noted that some landowners in Bishop and Big Pine had been contacted to join a suit against the city over pumping: "Barmore [one of the landowners] states that if the City will keep promises in connection with purchase of equipment and adjustment of reparation suit in Big Pine, he will be perfectly satisfied, otherwise feels that he and his crowd will have to join in suit."[59] In addition, to make sure that the point was not missed, the Grapevine siphon on the aqueduct was dynamited on November 11, 1931.[60]

Feeling the heat, General Manager of the Board of Water and Power Commissioners H. A. Van Norman, Assistant City Attorney W. Turney Fox, and Right of Way and Land Agent A. J. Ford sent a letter to the Senate Conservation Committee, saying that the city had employed a corps of valuation experts to value each town parcel with the advice and assistance of W. A. Avey, "a recognized expert in valuation matters and who was once Appraiser for the Banking Department of the State of California." Under these new appraisals, increases of over $1 million "were agreed to by the City because of its desire to establish harmony and facilitate cleaning up the Owens Valley situation." For those who still did not want to sell, the city offered to buy their surplus water rights and pay 25 percent of the value of the property and give the owner an option to sell for the remaining 75 percent within five years.[61]

While the board was trying to accommodate the demands of the town lot owners, it still had to be responsive to Los Angeles's water ratepayers and taxpayers. The conflicts may have been felt on the board, where there was some disagreement as to what to do about the continuing problems in Owens Valley and the intransigence of the town lot owners. Judge Harlan Palmer, appointed in 1929

TABLE 4.10

Town lot purchases by the City of Los Angeles, 1929–35

Town	Number of parcels	Total sales (dollars)	Mean sale price (dollars)
Bishop	459	2,947,698	6,422
Big Pine	169	905,164	5,356
Independence	183	766,038	4,186
Lone Pine	276	1,177,968	4,268
Laws	55	69,245	1,259
Total	1,142	4,866,113	4,261

SOURCE: Town lot values are provided in documents throughout the LADWP Archives. They have been collected and merged and are available from the author. Only observations that indicated the location of the property are included. Four observations in Manzanar and Sunland are not included.

by new mayor John C. Porter and who supported offering higher prices to the town residents as a means of clearing up the controversy, resigned as president of the board.[62] In all of this, there was a certain element of extortion at play. Under California legislation and the Supreme Court's ruling, Los Angeles had to compensate or buy the town properties, and the owners knew that. And by 1931 the board was boxed in with seemingly few options for settling the town lot issue.

Ultimately, a compromise was reached in 1931 and Los Angeles paid $5,798,780 to 824 owners for 1,300 town parcels, most of which brought little or no additional water to the city.[63] The prices paid were based on *1923* values that existed before major purchases by the city in the valley, and they did *not* reflect the 1929 agricultural depression that was affecting rural land values throughout the country.[64] Funds to buy town lots and remaining agricultural properties in Owens Valley required passage of a special 1930 bond election for $38.8 million.[65] Even so, some town property owners held out. The Johnson-Riley pool on Main Street in Bishop refused to sell until 1933. On February 21, 1933, B. E. Johnson and State Senator Joe Riley submitted a proposal to the LADWP offering to deliver twenty-seven properties involving eighteen separate owners for $344,867, as compared with the agency's 1929 price of $183,347 (a difference of $161,520). In 1927 the city had appraised the same lands at $172,015.[66] Eventually, most of the properties were acquired, and by 1935, Los Angeles had 95 percent of the agricultural acreage in Owens Valley and 88 percent of the town properties.[67]

Table 4.10 provides summary data for the sale prices received in the major towns in Owens Valley. The data are only for those town lots where there is information as to the location in the material in the LADWP Archives. Nevertheless, they cover most of the properties purchased by Los Angeles.

TABLE 4.11

Mean rural nonfarm home values in California and Nevada, 1930

California counties	Value (dollars)	Nevada counties	Value (dollars)
Inyo	4,127	Churchill	2,973
Lassen	2,248	Douglas	3,065
Los Angeles	4,850	Lyon	1,224
All California	3,659	All Nevada	<1,000

SOURCE: U.S. Census, *Population*, Volume VI, "Families, Reports by States, Giving Statistics for Families, Dwellings and Homes, by Counties, for Urban and Rural Areas and for Urban Places of 2,500 or More," Table 45, p. 38; California, Table 20, pp. 177–78; Nevada, Table 20, p. 824 (Washington, D.C.: Government Printing Office, 1933).

To understand the nature and the outcome of the exchange between the LADWP and property owners, it is useful to compare the mean sale prices in Table 4.10 with 1930 U.S. Census data on the value of rural nonfarm homes. The median value in the state of California was $3,659, and for the Great Basin state of Utah (probably the more appropriate state for comparison) it was $1,893.[68] In four of the five small rural towns in Owens Valley, the mean values exceed the census values reported for Utah by between $4,529 and $2,293, and for California by between $2,763 and $527, respectively. Of course, some of the properties in the five towns were commercial and therefore of higher value, raising the means. But the California statewide data are for areas of generally higher property values than in Owens Valley. The comparison with Utah census values shows gains to Owens Valley town residents of two to three times, seemingly more than adequate compensation for inclusion of commercial properties.

Table 4.11 provides more evidence on this issue with rural, nonfarm home values for counties in California and Nevada from the 1930 U.S. Census. As indicated in the table, the mean value of nonfarm rural homes in Inyo County is 13 percent greater than the California statewide median value, 84 percent greater than the values in comparable Lassen County, and only about 18 percent less than values in Los Angeles County. Inyo County values also far exceed those in similar Nevada counties. All in all it appears that town property owners did very well in their negotiations with the LADWP, securing prices that placed property values nearly on par with those in Los Angeles.

CONCLUSION

The Owens Valley story is one of conflict between the Los Angeles Board of Water and Power Commissioners and particular groups of farmers and town residents. At the time, the conflicts that took place in the 1920s were deemed California's "little Civil War." The analysis in this chapter explains why. First, there were high transaction costs of exchange exacerbated by the bargaining strategies adopted by the parties. Los Angeles sought the water-bearing lands in the valley and entered into sales negotiations with the farmers. For some, the exchange was quick and smooth. For others, especially those farmers with the most water, the exchange took time—up to ten years in some cases—and the negotiations were anything but smooth. Disagreements over valuation were complicated by bilateral monopoly conditions in the valley.

The farmers attempted to organize a single selling unit, the Owens Valley Irrigation District, to bargain with the LADWP. The board effectively countered this effort with the buyout of key properties. But eighty-six farmers formed three sellers' pools, and they held out for higher prices. All things equal, they did better in the land market, pushing up the prices they received by from $52 to $213 per acre, than the 228 unorganized farmers who were not on ditches and were scattered throughout Owens Valley. But their "cartel" was not strong enough to bring significantly higher prices for their water. The LADWP successfully used its bargaining power to price water more according to the value of its contribution to agricultural production than to the value of water in Los Angeles. On the margin, water in Owens Valley was not very valuable for agriculture, but it was very valuable in Los Angeles, and farmers wanted part of that extra value. Generally, however, they were not able to achieve it. The implicit battle over water values in bargaining over land is a critical and overlooked aspect of the Owens Valley story.

The issue of valuation and purchase of town lots also is a notorious part of the history of Owens Valley. Allegations of negative third-party effects of the purchase of farmland on commercial activities in the five towns galvanized opposition to the LADWP in Owens Valley; concerns about similar effects in current rural-to-urban water transfers are a critical part of opposition to water markets today. The nature of these third-party effects in Owens Valley also is examined in more detail in the following chapter, but here too, collusive action helped to raise the transaction costs of negotiation.

As with farmland, there were disputes over the valuation of town properties, including the evaluation of the proper characteristics and the relevant properties for baseline comparisons. An elaborate appraisal and price-fixing structure was set up involving representatives of both the towns and the LADWP. Even though the agency received State Supreme Court approval to purchase the town properties in 1929, and a large bond issue was passed in 1930 in order to fund those purchases, it took another five years of negotiations before sale agreements on many properties could be completed. Also, as with negotiations over farmland, leaders of sellers' pools in Bishop brought in the California legislature to bolster their bargaining position.

All in all, these were contentious negotiations, and in this case, where there was little or no water involved, the Owens Valley town residents appear to have done well. They were able to use 1923 property values as the base for any purchases of their properties, brought forward to 1929 values at an 8 percent growth rate. No other farming community in the Great Basin had such an opportunity, as indicated by relative census values for nonfarm rural homes.

5

THE GAINS OF EXCHANGE AND THE ORIGIN

OF THE NOTION OF "THEFT"

Farmers remain suspicious of the "Owens valley syndrome". . . . The "theft" of its water . . . in the early 20th century has become the most notorious water grab by any city anywhere . . . the whole experience has poisoned subsequent attempts to persuade farmers to trade their water to thirsty cities.

The Economist, *July 19, 2003*

Through the sellers' pools they organized in the 1920s into the early 1930s, Owens Valley farmers were able to raise the price they received per acre of land; farmers who were not in a pool and scattered throughout the valley received far less for their land. The mean price paid between 1916 and 1934 by the LADWP per acre of land in Owens Valley was $198, whereas the mean 1930 census per acre value of land and buildings in the four comparable Great Basin counties of Lassen in California and Churchill, Douglas, and Lyon in Nevada in 1930 was only $36.40. Although the agricultural depression had caused land prices throughout the region to decline, by any measure this difference in land values is striking. It is almost certainly due to the purchase of land in Inyo County by the LADWP.

The difference in farmland values suggests that Owens Valley farmers did well in their transactions with the board, better than if they had remained in agriculture, which was the determinant of land values elsewhere. The value of their properties on average was raised by about five times that of comparable lands in the other counties. Indeed, there are ample data to assess the overall impact of the water transfer from Owens Valley to Los Angeles, and the assessment is largely a positive one.

IMPACT ON THE TOWNS

Feared pecuniary effects on third parties play a major role in resistance to water markets today, and in Owens Valley town lot owners claimed that the city's actions to buy and export water from the region reduced the value of their

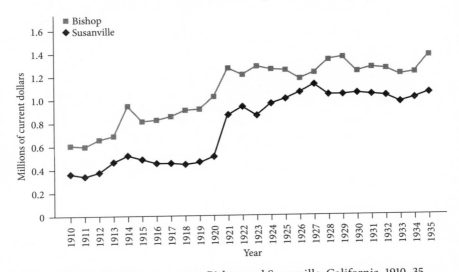

Figure 5.1   Municipal Assessments, Bishop and Susanville, California, 1910–35.

SOURCE: Report of the California State Board of Equalization, Sacramento: for 1910, Schedule G; 1911–14, Schedule H; 1915, 1916, Schedule G; 1917, 1918 Schedule F; 1919–22, Schedule D; 1923–34, Schedule E.

properties. They demanded compensation and eventually received very favorable buyouts from the Water Board in the early 1930s.

There are other data with which to assess the impact of the Water Board's actions on the value of property in the largest town, Bishop, and to compare that with a similar town, Susanville, the county seat of Lassen County in California, a similar Great Basin farming county that was affected by the agricultural depression of the 1920s but not by the purchase of water-bearing lands and export of water.

Each year the California State Board of Equalization presented the "Grand Total Value of All Property" (real estate, personal, and monetary) by municipality.[1] Figure 5.1 plots the data in current dollars for Bishop and Susanville from 1910 through 1935. The data do not reveal any deterioration in economic conditions as measured by property values in Bishop, either absolute or relative to Susanville, despite claims to the contrary. Property values in Bishop more than doubled between 1910 and 1921, rising from $607,920 to $1,269,580. With the advent of the agricultural depression that started with the collapse in commodity prices in 1921, property values rose more slowly in the 1920s, peaking at $1,355,686 in 1929. This nevertheless was a rise of 7 percent between 1921 and

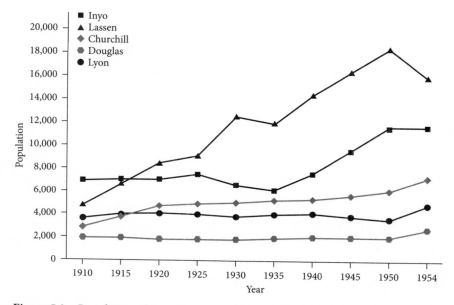

Figure 5.2 Population of Inyo County and Four Comparable Great Basin Counties, 1910–54.

SOURCE: For 1915, the population of all counties is interpolated from the 1910 and 1920 Decennial Census; Churchill, Douglas, and Lyon populations for 1915, 1925, 1935, 1945, and 1954 interpolated from Decennial Population Census; even years, U.S. Decennial Population Census; Inyo and Lassen Counties, 1925, from *The 32nd Report of the California Superintendent of Public Instruction for the School Years Ending June 30, 1925, 1926*; for 1935, *Report of the California Department of Education, June 1935*.

1929, when the actions of the Water Board were alleged to have been so damaging. In 1930, property values naturally declined with the Great Depression, but then rebounded to a level slightly above their 1929 values, at $1,358,390 by 1935. Susanville also had expansion over the period, buoyed in part by expansion of the lumber industry in Lassen County. Its property values, however, fell more steeply from their peak in 1930 through 1935. In any event, in both absolute and relative terms, Bishop's condition looks very favorable, at least according to State Equalization Board data on property values.

Another way to gauge the impact on the towns is to examine the population of Inyo County. If, as town lot owners argued, the local economy was deteriorating, the population should have been declining as labor migrated away. Figure 5.2 provides census population data for Inyo County, as well as for the four other Great Basin counties (Lassen, California; and Churchill, Douglas, and Lyon,

Nevada) between 1910 and 1954. These counties have similar agricultural characteristics common to the Great Basin: short growing seasons, relatively high transportation costs, alkaline soil, and limited rainfall. Livestock and alfalfa are the principal agricultural commodities. All four of the other counties retained their water, and two, Churchill and Lyon in Nevada, received additional irrigation water from the Newlands Project, constructed between 1903 and 1917. Lassen County, like Inyo, was denied a Bureau of Reclamation project.[2]

Inyo County's population growth was quite similar to that of the other four counties. It remained the second most populous county of the five, and virtually all of that population resided in Owens Valley. There is, however, evidence of a decline in population after 1925 through 1935, which would be consistent with claims that the local agricultural economy was disrupted. The population fell from 7,431 in 1925 to 6,212 in 1935. No other county experienced this large a population shift. After 1935, however, Inyo County's population had rebounded to 11,671 by 1954, an increase of 88 percent. This pattern of population growth is not consistent with a claim that the Owens Valley water transfer left the valley devastated and a desert.

IMPACT ON COUNTY PROPERTY VALUES

Returning to State Board of Equalization data, the figures regarding the total value of all property by county also are instructive. The data include the value of real estate (farm and nonfarm), improvements, personal property, money and solvent credits, and railroad assessments.[3] The comparative positions of property values in Inyo and Lassen Counties are shown in Figure 5.3.

Between 1900 and 1930, the total value of all property in Lassen County rose by 640 percent and in Inyo County by 917 percent. The value of all property in Inyo County grew by 43 percent more than in Lassen County during the thirty-year period of major Los Angeles land purchases in Owens Valley. Indeed, with the completion of the Los Angeles Aqueduct in 1913, Inyo property values moved past those in Lassen County, rising from $6,268,862 in 1912 ($7,431,405 in Lassen) to $9,505,223 in 1913 ($8,338,937 in Lassen). Inyo property values peaked in 1924 at $18,778,408 ($17,343,500 in Lassen) and then leveled off and declined slightly, until rising once again and peaking at $20,053,249 in 1931. Although Lassen did better from 1924 through 1931, property values fell more sharply in that county after 1931 through 1935.

The U.S. Population Census and California State Board of Equalization data, then, do not support the notion that the actions of the Water Board destroyed

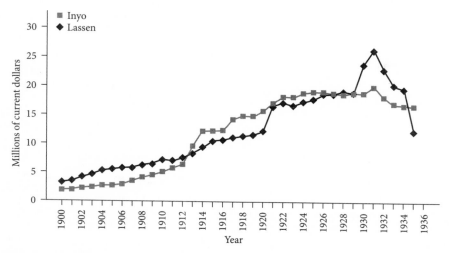

Figure 5.3   Property Values in Inyo and Lassen Counties, 1900–35.

SOURCE: *Report of the California State Board of Equalization*: for 1900–06, 1910–11, 1913, 1916–17, Schedule D; 1907–09, 1915, 1920, 1922, 1924, 1926, 1928, 1930, 1932, 1934, Schedule C; 1912, 1914, Schedule F; 1918, Schedule E; 1919, 1921, 1923, 1925, 1927, 1929, 1931, 1933, 1935, Schedule B.

the economy of Owens Valley (Inyo County) or of its largest town, Bishop. Although there was some decline in county population after 1925 through 1935, the value of property in both Bishop and Inyo County compared favorably in absolute terms as well as relative to the baseline town of Susanville and Lassen County. There was no deterioration in property values in Bishop, although there was a small falloff in Inyo County in 1932. Additional U.S. Census data on the agricultural economy of Owens Valley make it possible to assess the impact of the water transfer both absolutely and relative to the baseline Great Basin counties.

AGRICULTURAL LAND VALUES FOR
FIVE GREAT BASIN COUNTIES

Figure 5.4 illustrates the impact of Water Board purchases on land values in Inyo County relative to agricultural land values in the baseline counties between 1910 and 1954. The figure plots the value of agricultural land and buildings per acre. The runup in land prices in Inyo County during the 1920s, both in absolute terms and relative to the other counties, is very apparent. This was the period of

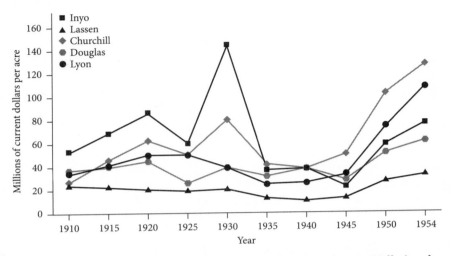

Figure 5.4   Land and Building Values per Acre, Inyo County (Owens Valley) and Comparison Counties, 1910–54.

SOURCE: Data assembled from the census by Barnard and Jones (1987, 10–12).

major Water Board purchases of water-bearing land in Owens Valley. It is also clear from the data that the change in land values in Inyo County after 1930 was probably not much different from that in the other counties. Los Angeles owned most of the farmland by 1930 and sold relatively small amounts after that, so the market would have been comparatively thin. Even so, the data reported in the census do not point to any destruction of land values in the valley.

Other census data help to make clearer the transitions that were under way in Owens Valley as the Water Board acquired the farmland there and to place these adjustments in the context of Great Basin agriculture. Figure 5.5 plots farm size in the five counties.

For the period 1910–25, Inyo had among the smallest farms in all five counties. But after 1930 the farms consolidated into larger units once purchased by Los Angeles. Farm size increased by 164 percent from 1930 through 1950, moving Inyo County farms into the mid-range of farm size across the five counties.

Figure 5.6 plots the number of farms for the five Great Basin counties. Just as farm sizes grew disproportionately in Inyo County, the number of farms fell more sharply there than in any other county, again reflecting the purchase and consolidation of farms by the Water Board. Except in Churchill and Lyon Counties, where the number of farms increased between 1910 and 1954, the number

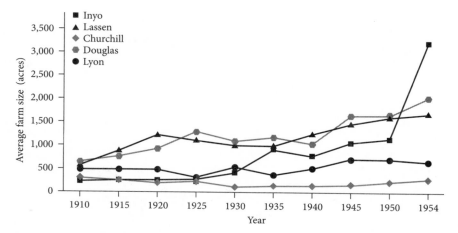

Figure 5.5 Farm Size in the Five Great Basin Counties, 1910–54.

SOURCE: Calculated from U.S. Agricultural Census 1910, 1920–54; 1915 data interpolated from the Decennial Census.

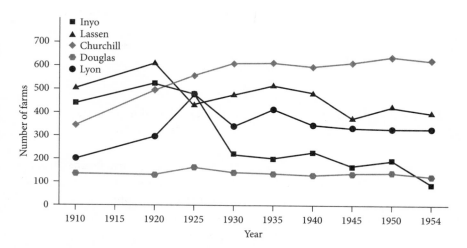

Figure 5.6 Number of Farms in the Five Great Basin Counties, 1910–54.

SOURCE: U.S. Agricultural Census 1910, 1920–54; 1915 data interpolated from the Decennial Census.

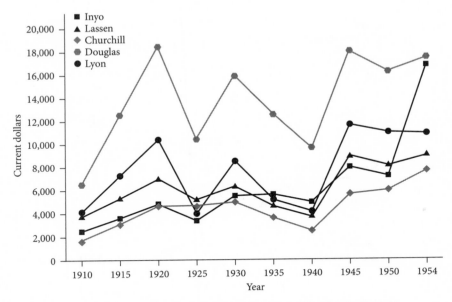

Figure 5.7   Value of Total Farm Production per Farm, Great Basin Counties, 1910–54.

S O U R C E : U.S. Agricultural Census 1910, 1920–54; 1915 data interpolated from the Decennial Census.

N O T E : Total Value of Farm Products = Value of Livestock Sold or Slaughtered + Value of Livestock Products + Value of Crop Production Divided by the Number of Farms.

of farms declined, and Inyo had by far the largest decline in units. Over that period, the number of farms fell by 6 percent in Douglas County, 21 percent in Lassen, and 74 percent in Inyo.

Figure 5.7 provides the total value of farm products per farm for the same five counties. In 1910, only Churchill had a lower value of production per farm than Inyo County farms, at $2,442 for Inyo and $1,621 for Churchill, whereas farms in the other counties were more productive: $3,777 for Lassen, $4,071 for Lyon, and $6,514 for Douglas. After 1910, ignoring the value of farm production in Douglas County, the value of production in Inyo County farms in current dollars rose through 1950, with a spike in 1954, in a manner similar to what is observed in three of the four baseline counties. Between 1910 and 1954 the value of production rose almost threefold, or 191 percent, to $7,100 in Inyo County. During the critical period 1920–35, when Los Angeles was acquiring farms in Owens Valley, the value of farm production per farm rose from $4,759 to $5,618. Again, there is no evidence that the farm economy was devastated by the Water Board's actions.

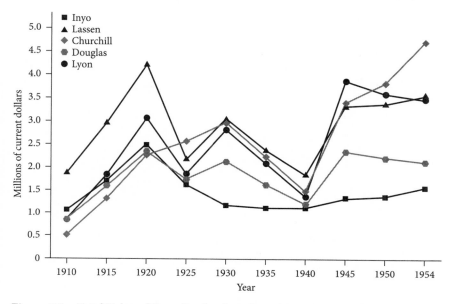

Figure 5.8   Total Value of Farm Production, Great Basin Counties, 1910–54.

SOURCE: U.S. Agricultural Census 1910, 1920–54; 1915 data interpolated from the Decennial Census.

NOTE: Total Value of Farm Production = Value of Livestock Sold or Slaughtered + Value of Livestock Production + Value of Crop Production.

The (now larger) farms that remained in production had greater values of output per farm.

Figure 5.8, however, which shows the total value of farm production in the five counties, does provide some evidence of a decline in the overall value of agricultural production in Inyo County. The total value of farm production as reported in the census peaked in Inyo County at $2,479,439 in 1920 and then declined through 1940, rising after that through 1954. Through most of the period after 1925, the value of agricultural production among the five counties was lowest in Inyo County.

Figure 5.9 reports the ratio of the value of livestock production to total farm production. Inyo County follows the other Great Basin counties in a move to livestock production, the comparative advantage of the region. In 1920 as Los Angeles was moving into the northern, agricultural part of the valley, the ratio of the value of livestock production to the total value of agricultural production was 42 percent in Inyo County. By 1954 the portion had risen to 97 percent, the

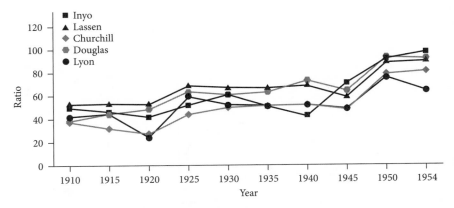

Figure 5.9 Ratio of Value of Livestock Production to Total Farm Production, Great Basin Counties, 1910–54.

SOURCE: Calculated from U.S. Agricultural Census data.

NOTE: Ratio = Value of Livestock Production + Value of Livestock Sold or Slaughtered Divided by the Total Value of Farm Production.

largest share for livestock of total farm production among the counties. As the Water Board purchased and consolidated farms in Owens Valley and diverted water from irrigation to urban use via the Los Angeles Aqueduct, the local economy shifted rapidly to livestock raising.

In total then, data from the U.S. Census and the State Board of Equalization provide the following assessment of the economy in Owens Valley during and after the purchase of water-bearing lands by the Los Angeles Water Board. The population of Inyo County dropped slightly from 1925 to 1935 and then rebounded and moved with that of the other Great Basin counties. The per acre price of land and buildings spiked in Inyo County relative to the other counties in the 1920s owing to the purchase of properties by the Water Board and then dropped back to the levels reported for the other counties. With the purchase of farms, average farm size in Inyo County rose along with farm sizes in the other counties. But the number of farms fell by more than elsewhere, consistent with a pattern of farm consolidation by the board. Inyo farms were among the smallest in the region in 1910, and therefore required the greatest consolidation in order to be economically viable over the long run. The value of farm production per farm rose, as did the share of livestock production, but the total value of farm production fell in Inyo County both absolutely and relative to the other counties. At the same time, however, total property values in Bishop and in Inyo County rose.

These data describe a rural economy in transition from small farms with some irrigated crop production to larger farms specializing primarily in livestock. This transition would have happened with or without the acquisition of farms in Owens Valley by Los Angeles. Los Angeles speeded the process of transition with its wholesale purchase of small farms in the 1920s. Its introduction of a support labor force for development and maintenance of the aqueduct and other water and hydro power infrastructure also served to bolster the value of property in Inyo County and in the county seat of Bishop. The growth of Los Angeles and improvements in transportation also drew greater numbers of visitors to the region. As the number of farms declined, farm size grew and the value of individual farm production rose. Some population migrated from the valley, but by 1935 the valley's population was again rising. There is no evidence of a collapse or serious economic deterioration in the county from the water transfer, contrary to the common contemporary view.[4]

The rise in per acre property values also indicates some benefits to the existing population that would not have been available had the farms been purchased by other farmers for consolidation. Los Angeles paid more. The runup in prices in Inyo County between 1910 and 1930 was not matched in any of the other counties. In 1930, per farm and per acre land values in Inyo County were $62,200 and $143, respectively, up almost fivefold from 1910 for farms and almost threefold for acreage. No other county had anything close to this increase in value.[5] This price pattern reflects the purchases of farm properties by the Water Board because there are no other indications of productivity change. This evidence also supports the claim of the board that it was paying well above market rates for Owens Valley properties. These mean values also reflect the effects of the successful holdout strategies of the sellers' pools.

There are other ways of assessing the impact of the sale of property to the Water Board on farmers in Owens Valley. Between 1900 and 1930, census land values in Owens Valley rose by around a factor of eleven, increasing from an average of $13 per acre to $143.[6] By contrast, land values in the baseline Lassen County rose by a modest two times over the same thirty-year period, from $10 per acre to $21. These data suggest that most of the rise in land values in Inyo County was due to land purchases by Los Angeles and not due to changes in agricultural commodity and livestock prices.

At the same time, the value of agricultural land and buildings in Inyo County rose by $11,757,724, from $1,801,810 to $13,559,534, an increase of 653 percent. By contrast, farm property values in Lassen County increased by $6,306,099, from

$3,657,520 to $9,963,619, or 172 percent from 1900.[7] Again, the baseline Great Basin county did not do as well.

An alternative way of assessing the impact of Owens Valley land sales is to consider the counterfactual of no Los Angeles purchase or export of Owens Valley water, the expansion of farm acreage in Inyo County at the same rate as occurred in Lassen County, and the same increase in land prices in Inyo as occurred in Lassen. Under this plausible counterfactual, farmland values would have been $4,547,738 in 1930 in Owens Valley.[8] But this value is over $9 million *less* than what actually is observed.

Regardless of how the gains are measured, if one uses Lassen County and the other Great Basin counties as a baseline, Owens Valley landowners did better on average by selling to Los Angeles than by remaining in irrigated agriculture. Owens Valley landowners captured part of the aggregate gains of trade, as did property owners in Los Angeles. These data are indicative of the dramatic size of the aggregate benefits of this early water exchange, even when none of the increase in urban land values in Los Angeles is included.

In sum, the U.S. Census and California Board of Equalization data suggest that bargaining produced a more positive outcome for Owens Valley landowners than is commonly believed. The export of water reduced crop production as a share of overall agricultural output and encouraged a shift toward livestock. But this pattern also took place in the other counties. The comparative advantage of the Great Basin ultimately was in livestock, so there would have been a gradual shift from crops in Owens Valley, even had the aqueduct not been built. Owens Valley was not left a wasteland, as is sometimes alleged. Nor would its small orchards and other crops that were grown through 1920 likely have remained competitive for the longer term had the water remained in the valley. The export of water did change agriculture and life in Owens Valley, but it was not dominantly a negative change; nor was it decidedly different from that which occurred in agriculture throughout the Great Basin.

THE GAINS TO LOS ANGELES

Although Owens Valley farmers did better by selling their farms to Los Angeles than if they had remained in Great Basin agriculture, property owners in the city captured a much larger share of the gains from trade. The water market analysis in the previous chapter indicates that the Water Board was able to use its market power to pay less for Owens Valley water than it might have been willing to pay. The board was able to pay a price that was closer to its agricultural value than to

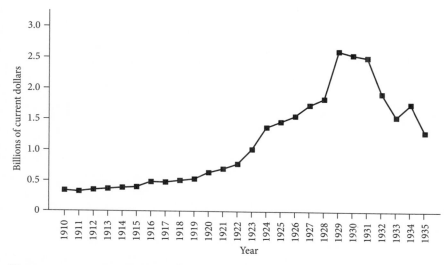

Figure 5.10   Los Angeles City Property Values, 1910–35.

SOURCE: *Report of the California State Board of Equalization*, Sacramento: for 1910, Schedule G; 1911–14, Schedule H; 1915–16, Schedule G; 1917–18, Schedule F; 1919–22, Schedule D; 1923–34, Schedule E.

its value in Los Angeles. Given the problem of "excess water" in the valley, where much of it had low marginal farm value, the board was able to keep the prices it implicitly paid for water low relative to Los Angeles values. This is the basis for the notion of "water theft"—not that the city literally "stole" the water, but that it got the water for less than it might have been willing to pay.

There are other indications of the magnitude of the benefits received by the city. According to U.S. Census data, between 1900 and 1930 the value of agricultural land and buildings in Los Angeles County rose by $406,451,090, from $70,891,930 to $477,343,026, an increase of 673 percent. While property values rose from 640 to 917 percent in Lassen and Inyo Counties between 1900 and 1930, the total value of all property in Los Angeles County rose by *4,408* percent, to $26,553,158,282 from $100,136,070. Figures 5.10 and 5.11 report the value of all property in the city and county of Los Angeles using State Board of Equalization data.

The reported increases in property values in Los Angeles are astounding, and they depended on a steady supply of Owens Valley water. The only alternative source was the Colorado River, which did not arrive until 1941. In the absence of Owens Valley water, Los Angeles would have been a much smaller community with less political influence. It is quite possible that under this scenario, Hoover Dam and the Colorado River aqueduct system would not have been built, or

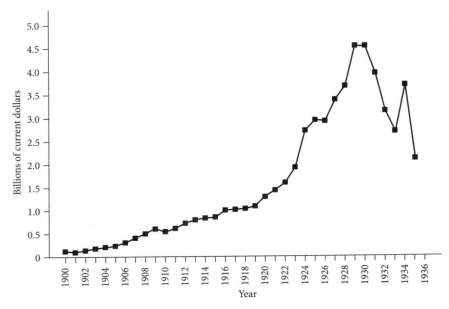

Figure 5.11 Los Angeles County Property Values, 1900–35.

SOURCE: *Report of the California State Board of Equalization*, 1900–06; for 1910–11, 1913, 1916–17, Schedule D; 1907–09, 1915, 1920, 1922, 1924, 1926, 1928, 1930, 1932, 1934, Schedule C; 1912, 1914, Schedule F; 1918, Schedule E; 1919, 1921, 1923, 1925, 1927, 1929, 1931, 1933, 1935, Schedule B.

at least not to the size that they were. Given the magnitudes of the increases in property values in Los Angeles, it is no wonder that Owens Valley farmers wanted a greater share of the benefits of reallocating their water.

CONCLUSION

The Owens Valley water transfer to Los Angeles was not the disaster asserted and repeated in the contemporary literature and the press. Owens Valley was not left a wasteland or destroyed at the time; nor is it a wasteland today, as anyone who drives through the valley can attest. Indeed, under Los Angeles's steward-ship, the valley has a pristine, undeveloped, natural character that is not found in most parts of the southeastern Sierras, where urban sprawl is much more characteristic. The small towns are more economically vibrant than many of the other communities that dot the Great Basin, which are just barely surviving. In the 1920s, the farmers and townspeople of Owens Valley did not have their properties, especially their water, literally stolen. Rather, they were able to do very well in their land negotiations with the Los Angeles Water Board. They certainly did much better than if Owens Valley had been left alone to gradually lose its agricultural base, as was the common situation throughout the Great Basin.

The region is not especially good for agriculture, particularly the small-farm agriculture that characterized Owens Valley in the early twentieth century. Because most Inyo County farms were smaller and had lower production per farm than those elsewhere in the region, the adjustments necessary to achieve a more sustainable farm economy would have been even more severe than in other places. Farm consolidation and out-migration of the agricultural labor force were necessary. Although these occurred after Los Angeles bought the valley's farms, the impact was mitigated by other Los Angeles–related activities. The city's investments in highways linked Owens Valley to the more populous south for tourism. And the aqueduct and other water infrastructure required maintenance and an associated labor force to be resident in the valley.

All told, Owens Valley did well in comparison with other Great Basin regions. Indeed, it does not deserve to be cited as the leading example of the dangers of water exchanges from agriculture to urban and environmental uses. Its outcome was favorable for the parties involved and should be presented as such. The conflict (and resulting allegations of water "theft") was over the distribution of the gains from trade in which both parties participated.[9]

But there was an imbalance in that distribution, and the negotiations over lands were long and arduous with extensive coverage by the press and the intervention of the governor and the California legislature. The aggregate gains to Los Angeles landowners were forty times or more greater than those of Owens Valley residents from the redistribution of water through its sale.[10] The perception of unfairness over the terms of trade also was driven by the nature of supply and demand for water. Urban users had relatively inelastic demand, whereas farmers competing for sale had a comparatively elastic export supply. Hence Los Angeles residents gained considerable consumer surplus from the transaction.[11] The effort of farmers to reap more of the gains of trade in negotiation explains the formation and relative greater success of the sellers' pools. Even so, a disproportionate share of the returns went to Los Angeles.

The experience of Owens Valley emphasizes the importance of resolving distributional conflicts in water transfers, where the amounts of money at stake can be very large. So much water is placed in low-value uses at the margin in agriculture, and urban and environmental demand prices are so much higher. These disputes play an important role in hindering current water transfer efforts, especially as they apply to alleged third-party effects in local farm communities. The tortuous record of negotiations in Owens Valley, despite large *ex post* aggregate gains from trade, highlights the importance of addressing these disagreements in order to smooth water transfers. Given the likely surpluses generated from the reallocation of water, the basis for addressing such concerns seems to be at hand. The allocative benefits will swamp distributional concerns.[12]

Owens Valley also resonates in the minds of many because of the dust pollution that was generated from Owens Lake, which was made dry (or at least more often and more completely dry) by the diversion of Owens River water to Los Angeles via the aqueduct. To many observers, the diversion of water from feeder streams in the Mono Basin to the north of Owens Valley (where Los Angeles went in the 1930s in search of more water) and the associated effects on the declining level of Mono Lake is another example of the rapaciousness of Los Angeles. These environmental issues and their resolution are examined in the following chapters as continuations of the "legacy of Owens Valley."

6

WATER RIGHTS AND WATER REALLOCATION
IN OWENS VALLEY, 1935–2006

> It being quite obvious that at the present rate of growth of [Los Angeles], an
> increased supply of water will be needed long before the estimated time of ten
> years, allowed in conservative estimates for the completion of the Colorado
> River project. . . . There is available to the City at this time an additional water
> supply in the two watersheds of the Owens River and Mono Basins. In the
> Owens River drainage, from which the City is now drawing what is contained
> in the Aqueduct, your engineers estimate there is a safe supply of 385 second
> feet. Mono Basin contains a supply of approximately 235 second feet.
>
> *Letter from Los Angeles mayor John C. Porter to members of the Board
> of Water and Power Commissioners, 1929*

We have seen how the efforts of the parties to obtain more of the benefits of exporting water to Los Angeles raised the transaction costs of exchange and fixed perceptions of the fairness of the trade. Indeed, although farmers on average did well in selling their properties to Los Angeles, one legacy of the bitter negotiations that took place was the notion, especially held later in the twentieth century, that the rights acquired by the city were somehow not legitimate, that they had been stolen. This perception may have helped fuel the subsequent controversial history of control over Owens Valley water.

This and the following two chapters shift the discussion to the experience of Los Angeles's water rights in Owens Valley and the Mono Basin during the period 1935 through 2006. The analysis relies less on quantitative evidence and more on qualitative information drawn from environmental impact reports, judicial rulings, regulatory proceedings, and the secondary literature, including law reviews. There is a large literature on the battle between Los Angeles and environmental groups over the export of Mono Lake water after 1970 and a much smaller one on the conflict between the city and residents of Owens Valley over management of source waters.[1] The contribution here is to assess the likely costs of the continuing litigation that took place between the parties and to suggest alternative approaches for similar water conflicts that may take place elsewhere.

In the absence of hard data, there necessarily is more speculation and opinion in these chapters. Nevertheless, the key arguments are as follows. When Los Angeles acquired water rights in Owens Valley and the Mono Basin, urban use and economic development were viewed as the highest value for the water. This was typified in President Roosevelt's statement in 1906 when he claimed, "It is a hundred or thousand fold more important to the State and more valuable to the people as a whole" if the water were used in Los Angeles than in Owens Valley.[2] Wildlife habitat, recreation, and amenity values in the source region were considered of much lower worth. Accordingly, any physical externalities from the export of water would have been of relatively little concern had they been anticipated.

Over time, however, these values have risen, and they have been affected by the shipment of water by Los Angeles from Owens Valley and the Mono Basin. Irrigation water was reduced, constraining agriculture and the rural community that both residents and visitors sought to protect. Groundwater pumping increased, especially in the 1930s during drought and after 1970 to fill the second aqueduct. This action not only further reduced agricultural water; it also damaged riparian and aquatic habitats. Trees and other less drought-tolerant plants died. The scenic beauty of the region was altered to some degree. Diversion of water dried up Owens Lake, increasing dust pollution from its bed, as well as the streams tributary to Mono Lake. Fisheries were lost in the Owens River and in Rush, Lee Vining, Walker, and Parker Creeks. Mono Lake's level dropped dramatically, threatening a major bird flyway (for California Gulls) as well as the visual value of the lake.

Some of these effects were internalized, at least in part, by Los Angeles's ownership of most of the land. Other effects were less so. For the latter, addressing the physical externalities required limitations on the export of water. A key question is, what is the most efficient way of addressing these problems? Here, efficiency considerations include the costs of responding to the externality (including the opportunity cost of the water) and the transaction costs of agreeing to do so; the timeliness of the resolution; whether the reallocation of water meets cost-benefit standards; and whether the security of water rights is maintained so as to encourage investment and trade in water and its infrastructure throughout the West. Actions taken in regard to Owens Valley and the Mono Basin set precedents for similar water problems elsewhere. Another question is, which externalities merit regulatory intervention?

There are no easy answers. There are many claimants and values associated with water. In a semi-arid region, the issues are sensitive ones, and a great deal is at stake in effectively responding to these problems. This and the next two chapters outline the history of judicial, regulatory, and political conflict over Owens Valley and Mono Basin water, whereby Los Angeles gradually lost rights to it. The process was hotly contested by all parties. The experience is compared with

TABLE 6.1
Owens Valley farm- and ranchland acquisitions
by the city of Los Angeles, 1905–45

Years	Acreage
1905–22	100,838
1923–29	128,190
1929–32	39,972
1932–34	936
1934–45	8,119

SOURCE: "Chronological Statement of Some Facts Pertaining to Land, Construction, Water Supply and Organization Matters of the Department of Water and Power in the Owens Valley District from December 10, 1928, to February 1, 1945, including a General Statement of Facts from 1895 to December 9, 1928," February 1, 1945, Bibliography file, Tape EJ00087, LADWP Archives.

a hypothetical alternative of greater reliance, instead, upon the negotiated purchase of Los Angeles's water rights by private parties where possible or by government agencies where there were significant public goods issues at stake.

Compensation for water acquired in the past by Los Angeles to address externalities involving new environmental and recreational values is not strictly an issue of distribution. Rather, it seems likely that it would have had worthwhile efficiency consequences in speeding the response to the physical externalities of water export, in reducing transaction costs, and in generating something closer to an "optimal" water allocation. It is not possible to test the claims made here, but the record suggests that the litigious methods actually used were extremely contentious, slow, and costly, and that the information generated by the competing parties about relative water allocations may have been too biased to provide accurate guidance for the weighing of alternative values.

THE MANAGEMENT OF CITY PROPERTY RIGHTS TO LAND AND WATER IN OWENS VALLEY, 1930–2006

By January 1, 1938, Los Angeles owned almost 99 percent of the ranchland and 88 percent of the town and commercial lot acreage in Owens Valley.[3] Eventually, as it acquired more properties in the Mono Basin, Los Angeles came to own some 306,000 acres of land in Inyo and Mono Counties. These are surely the largest landholdings of any local government in the United States.[4] Table 6.1 summarizes the purchase of ranchlands in Owens Valley between 1905 and 1945 by the Los Angeles Water Board and illustrates the concentration of city acquisitions before 1932.

As Los Angeles obtained farmland, it consolidated units into larger ranches. This pattern is evident in Figures 5.5 and 5.6, which show that in 1925 Inyo County had the largest number but the smallest-sized farms among the five comparable Great Basin counties—Inyo and Lassen, California; Churchill, Douglas, and Lyon, Nevada. By the 1930s, however, the number of Inyo County farms had declined sharply and farm size had increased. As a result of this farm consolidation, as well as the lure of greater economic opportunities available in Los Angeles and other urban areas, many members of the local population migrated out of the valley. Indeed, there was a comparative decline in the Inyo County population between 1925 and 1935. Those who remained leased their farms or their town properties from the LADWP.

Its leasing policies gave first consideration to those who had previously lived on the properties. Five-year renewable leases were granted, and irrigation was allowed so long as there was excess water beyond the requirements necessary to fill the aqueduct. This was usually the case before 1970. By that year, as Los Angeles's population reached 2,816,061, Owens Valley was supplying about 80 percent of the city's water, plus the aqueduct flow was generating electricity.[5]

Even so, there was growing contention between the LADWP and Owens Valley residents over the city's water rights and water policies as time passed. These conflicts escalated after 1970 with the completion of the second aqueduct and the ability of the city to export more water from the region. The disputes were interrelated, and they centered on the amount of water for local irrigation water, the extent of groundwater pumping by LADWP, the transfer of town property ownership, taxation of LADWP property, and the diversion of more water to protect aquatic and riparian habitat, as well as to increase stream flows and partially refill Owens Lake. Gradually, as the conflicts became more political and intense, judicial and administrative rulings forced Los Angeles to release water from the aqueduct, to engage in costly capital investments to mitigate dust pollution from the Owens Lakebed and rehabilitate aquatic and riparian habitats, and to submit to greater county, state, and federal regulation of its water management policies in Owens Valley.

The most intense local disputes arose over the quantity and security of irrigation water made available for agriculture and the magnitude of groundwater pumping used by the city to augment surface water supplies for the aqueduct. With regard to irrigation water, the most commonly referenced work on Owens Valley, by William Kahrl, claims that the city's leasing policies forced abandonment of once-productive farmlands because leases were short-term and insecure, with cancellations during drought years.[6] But as noted earlier, Owens Valley was never very productive agriculturally, and the lessees knew the conditions of the leases that stipulated water supplies were subject to aqueduct needs.

Indeed, except for irrigation cancellations during the drought of 1930, water was available for farming (mostly alfalfa and pastures), frequently reaching 30,000 acres through 1967. Usually, Owens Valley farmers could be generous in their water use, employing flood irrigation and applying up to 10 acre-feet of water per acre as compared to 3 to 6 acre-feet per acre or less in contemporary California agriculture.[7] Between 1945 and 1967, there were strict limits on irrigation six times, or one every three to four years, when very dry conditions led the board to curtail water for irrigation so that more could be sent down the aqueduct.[8]

It is understandable that lessees would have been unhappy about any reduction in their water allocations, but they had contracts with lower-priority uses. This same problem occurs throughout the West under the appropriative water rights system, whereby senior rights holders have first claim on surface water before the demands of lower-priority claimants are met. During drought, when flows are diminished, low-priority rights holders often do without or acquire water from their neighbors through short-term leases.

In addition, during dry years, when the board reduced irrigation water, it also increased groundwater pumping. Owens Valley is rich in groundwater, with much of it close to the surface. The groundwater basin has 30 million acre-feet of water or more within 1,000 feet of the surface. The basin also is extensive, running 120 miles by 15 to 30 miles from north of Bishop to south of the Haiwee Reservoir on the Los Angeles Aqueduct.[9] As the surface landowner, Los Angeles had a right to access the water below its properties, so long as there were no negative collateral effects on adjacent properties. Since the city owned virtually all of the land, these external effects of pumping on land values would have been minimized. Even so, as groundwater levels fell and surface watercourses that were in communication with groundwater dried up, aquatic and riparian habitats were damaged. Fisheries, other wildlife, and plant life were affected, as was the visual quality of the landscape.

Owens Valley aquifers were stores of water, and they provided balance for the aqueduct when surface runoff was insufficient. During the drought years of 1929–32 the Los Angeles Water Board extracted 336,000 acre-feet of groundwater from around Bishop and Independence, and in 1931 the city's wells were producing 30 percent of the total aqueduct supply. This accelerated pumping brought criticism and helped lead to adoption of the 1931 County of Origin Law in California, which prohibited export of water from one area to another for

TABLE 6.2

Los Angeles Aqueduct water sources from Owens Valley, 1940–89 (in acre-feet)

Year	Owens Valley surface runoff	Owens Valley ground-water	Aqueduct flow to Los Angeles	Year	Owens Valley surface runoff	Owens Valley ground-water	Aqueduct flow to Los Angeles
1940	441,898	0	204,557	1966	349,868	603	319,922
1941	648,446	0	243,452	1967	684,936	1,325	340,069
1942	522,933	131	264,215	1968	342,457	23,914	345,846
1943	499,464	52	271,987	1969	973,729	4,023	343,767
1944	404,037	0	274,004	1970	407,147	34,285	437,413
1945	590,626	11	290,725	1971	363,032	149,562	471,525
1946	487,526	34	293,932	1972	324,924	173,008	452,019
1947	366,550	52	314,136	1973	519,584	86,751	455,347
1948	297,689	25	283,747	1974	488,837	78,647	465,176
1949	338,077	27	300,838	1975	404,157	116,924	478,066
1950	356,596	23	310,900	1976	291,586	118,579	399,627
1951	368,859	27	317,243	1977	255,417	153,024	288,640
1952	644,955	29	314,510	1978	686,471	43,207	507,285
1953	361,655	24	313,629	1979	445,747	96,871	482,781
1954	377,571	27	316,589	1980	683,870	45,144	499,167
1955	380,713	28	320,243	1981	403,773	108,400	464,125
1956	539,761	33	318,723	1982	751,937	45,884	503,049
1957	423,291	27	328,048	1983	881,746	44,922	534,113
1958	588,789	26	319,975	1984	560,386	61,981	506,587
1959	310,706	38	323,403	1985	480,287	107,718	498,574
1960	260,242	65,096	320,001	1986	711,372	69,887	506,041
1961	253,608	109,985	333,030	1987	328,682	209,394	426,908
1962	482,085	9,175	326,583	1988	304,064	200,443	360,208
1963	504,072	899	334,340	1989	302,349	155,903	240,152
1964	307,125	25,100	332,963	Average	463,330	46,962	364,360
1965	462,845	6,732	319,839				

SOURCE: Jones and Stokes (1993, 3A-6), and for Mono flows, personal communication with Paul Scantlin, Civil Engineer, Water Department, LADWP. These include monthly water exports through Mono Craters Tunnel. Monthly data were summed for January–December for each year.

NOTE: Aqueduct flow totals vary somewhat across sources, most likely due to differences in months included and in conversion factors.

economic development.[10] When the drought ended by the middle 1930s, pumping dwindled. Indeed, between 1933 and 1963, groundwater pumping averaged only 10 cubic feet per second (cfs), or 7,227 acre-feet a year, and in 1940, Los Angeles gave up the right to pump from a 98-square-mile area southwest of Bishop where most of the region's population was concentrated.[11] This action was to reduce any externalities in the most populous area, where most of the remaining private property existed.

Table 6.2 provides data on the various Owens Valley sources of Los Angeles Aqueduct water from 1940 through 1989. The figures reveal the relatively limited sustained reliance on pumping until the second aqueduct was opened in June 1970.

THE SECOND AQUEDUCT AND CHALLENGES
TO THE CITY'S WATER RIGHTS: AGRICULTURAL
WATER, GROUNDWATER PUMPING, AND
ENVIRONMENTAL CLAIMS

The second aqueduct from Owens Valley to Los Angeles was completed in 1970, and it increased total flow capacity by 290 cfs, or by 33 percent from the first aqueduct's capability of 485 cfs at the time.[12] To take advantage of its new ability to export water, the LADWP planned to increase groundwater pumping, divert more of its water from the Mono Basin, and reduce agricultural irrigation to 15,000 acres.[13] The Mono extension of the aqueduct, completed in 1940, had never been used to its potential because the original aqueduct was not large enough to carry all of the water to Los Angeles. When the second aqueduct was completed, the Mono extension was projected to provide about 123 cfs annually (89,000 acre-feet) on average to the city. Further, the LADWP's 450 wells could add between 146 cfs and 250 cfs each year.[14]

There was, however, sharp reaction in Owens Valley to the city's plans to increase pumping and to reduce irrigation water to fill the aqueduct. Instead, Inyo County demanded more water, not less, for local agriculture, recreation (reservoirs and stream flows), and riparian habitat. There were fears that the proposed groundwater extraction would damage local vegetation near the wells and harm the recreation-based economy in the valley.[15]

In response, the LADWP agreed to retain some additional water for irrigation and recreation, but the agency was unwilling to compromise its overall ability to fill the two aqueducts. Indeed, in the face of a very severe drought in the early 1970s, the LADWP began consideration of a third aqueduct, intensifying concerns in the valley about the loss of water. On August 20, 1972, the Inyo County Board of Supervisors rejected the city's water export plans and demanded limits on groundwater pumping, maintenance of minimum and maximum flows in the Owens River, and abandonment of proposals for a third aqueduct.[16]

Following up, Inyo County filed suit under the California Environmental Quality Act of 1970 (CEQA) on November 15, 1972, contending that the LADWP was required to assess the environmental impact of its pumping program with an Environmental Impact Report (EIR) before increased groundwater extraction could take place.[17] The agency countered by arguing that the pumping program was designed to fill the second aqueduct, which was authorized, constructed, and operating before the advent of the law. According to the argument, the second aqueduct and the pumping necessary to fill it were part of a single, integrated operation that predated the CEQA.

This court battle was the beginning of *thirty-four years of litigation, constituent group lobbying, and regulatory interventions* over control of Owens Valley and Mono Basin water. Writing as early as 1980, before most of the clashes had taken place, William Kahrl labeled these battles "the Second Owens Valley War."[18]

Initially, the LADWP was able to secure judicial support for its contentions. The Sacramento County Superior Court agreed with the LADWP and denied Inyo County's demand for an injunction against groundwater extraction pending completion of an Environmental Impact Report. But on appeal June 5, 1973, in *County of Inyo v. Yorty*, 32 Cal. App. 3d 795 (1973), the Third District Court of Appeals in Sacramento ruled that the California Environmental Quality Act indeed applied.

The appeals court separated plans to supply the second aqueduct with water from its earlier construction and operation. While the LADWP was preparing the mandated EIR on its plan to accelerate subsurface water withdrawals, pumping was limited by the court to the quantity allowed on November 23, 1970, the date that the CEQA became operative. As a result of this and subsequent court rulings, between 1973 and 1984 groundwater pumping by the LADWP was restricted to about 149 cfs, well below the maximum the city's wells could provide.

The LADWP issued a Draft Environmental Impact Report on August 29, 1974. Inyo County rejected its conclusions and demanded a new evaluation. Absent agreement, the restriction on pumping continued. Faced with reduced groundwater sources for the aqueduct, the LADWP cut water on its irrigation and recreational leases in Owens Valley on September 20, 1974. Although the agency had the right to do this and had a mandate to maintain aqueduct flow, the residents of Owens Valley viewed the action as retribution, and it served to sour relations even more. The Sacramento County Superior Court, however, intervened and ordered Los Angeles to restore water deliveries while it prepared a new EIR.[19]

The drought of 1975 through 1977 brought some relaxation in the court's restrictions on pumping so that aqueduct flows could be sustained. But in 1976, Inyo County was joined by other constituent groups, including environmental organizations such as the Sierra Club, in emphasizing not only the negative effects on vegetation and riparian habitat of groundwater withdrawals, but also the problem of air pollution from dust blowing off the dry Owens Lakebed.[20]

The new, three-volume final EIR was issued in May 1976 by the LADWP (*Final Environmental Impact Report on Increased Pumping of the Owens Valley*

Groundwater Basin). As before, Inyo County on August 17 challenged the conclusions and the water management plan outlined in the document. While the
court evaluated the respective claims and allowed higher levels of pumping for
the duration of the drought, it required that the city maintain agricultural irrigation deliveries at three-fourths of normal.

This requirement seems questionable given the low value of the agricultural
use of water in Owens Valley, while Los Angeles residents were faced with mandatory water rationing.[21] It demonstrates, however, how politicized and emotional
water allocation and use became as the disputes intensified. Indeed, the LADWP
countered with efforts to meter water use in the Owens Valley towns as well and
to raise water rates there. Inyo County successfully resisted these plans.[22]

In *County of Inyo v. City of Los Angeles*, 71 Cal. App. 3d 185 (1977), the appeals
court dismissed the second Environmental Impact Report as being legally insufficient. The city appealed the ruling to the California State Supreme Court, but
on October 6, 1977, the court rejected the appeal without comment. Groundwater pumping and water allocations within Owens Valley remained regulated by
the court at 149 cfs. An attempt by the LADWP to get a third EIR accepted also
was refused by the court in August 1978.

As the litigation was taking place, the Los Angeles Aqueduct suffered damage
by dynamite, and other city property was vandalized. There also was agitation
for the sale of city-owned town properties (which Los Angeles had purchased in
the late 1920s and early 1930s to address third-party effects of water export) and
for higher county taxes on Los Angeles–owned assets in Inyo County. Numerous lawsuits over the county's assessments and tax policies followed.[23]

Los Angeles sold some of its properties earlier in the 1940s and 1950s, and
it began selling additional units in 1979. There was risk to the city from these
transactions, however, because the more private land in Owens Valley the less
the city internalized any negative third-party effects from its water management
policies. The LADWP also resisted property sales and associated economic development in the valley because those activities could constrain its options for
supplying water to the aqueduct, as already was becoming apparent.[24] Further,
as by far the largest land owner in the county and with fixed assets that could
not be moved, the city was susceptible to higher taxes as a punitive measure
from Inyo County, which was embroiled in conflict with the LADWP over water
rights and allocation.

Inyo County, the Concerned Citizens of Owens Valley, and the LADWP
sought a compromise to the impasse over preparation of the Environmental
Impact Report with an agreement for a five-year joint study of the effects of

groundwater pumping in 1979. The parties could not cooperate, however, and the arrangement fell apart in 1980. Inyo County followed with a new suit against the city demanding a revised EIR that complied with the CEQA, and the court concurred in *County Inyo v. City of Los Angeles* 124 Cal. App. 3d 1 (1981).[25]

In the meantime, Inyo County moved independently to regulate the city's groundwater extraction. In April 1980 an "Ordinance to Regulate the Extraction of Groundwater Within the Owens Valley Groundwater Basin" was released for review by the Inyo County Board of Supervisors. Local constituent groups lined up support, and it passed overwhelmingly on November 4, 1980. The ordinance called for a water management plan that maintained the groundwater table at a depth that would support natural vegetation and wildlife and that would minimize air pollution from blowing dust. It created the Inyo County Water Department and Inyo Water Commission to issue pumping permits and to administer the management plan. It also authorized the Inyo County Board of Supervisors to impose fees on pumping.

Had the ordinance been sustained, it would have sharply redefined the city's vested groundwater rights. Los Angeles successfully challenged the constitutionality of the ordinance, however, and it was rejected by the Superior Court in July 1983.[26] A similar ordinance was adopted by the county fifteen years later on July 27, 1999.[27] By that time the water rights held by the LADWP in Owens Valley already had been significantly modified.

Meanwhile, in 1982 the LADWP and Inyo County adopted a Memorandum of Understanding (MOU) for groundwater management that outlined the objectives of both parties. A joint Los Angeles/Inyo Technical Group was created to study water conditions in the valley funded by the LADWP, and the U.S. Geological Survey began an evaluation of the valley's hydrology and groundwater extraction. The agency's simulations suggested that a 50 percent reduction in groundwater withdrawals would be necessary to maintain the water table at targeted levels and to preserve biomass.[28] Alternative sources of water for the aqueduct were not outlined.

The MOU subsequently would be characterized more by a *lack* of understanding than a lack of cooperation. The judicial and administrative process under way offered few options for contracting between the parties. Los Angeles's water rights were not accepted as a basis for exchange.

On September 28, 1984, after over eleven years of costly conflict, the court offered to modify its pumping restrictions if the parties could reach lasting agreement on a water management plan (*County of Inyo v. City of Los Angeles*, 160 Cal.

App. 3d 1178). As a result, the city and the county concluded a five-year "Interim Agreement" in 1984. In the agreement the parties: (a) settled existing property tax litigation; (b) temporarily suspended Inyo County's appeal of the ruling that invalidated its groundwater ordinance; (c) developed a long-term groundwater management plan; (d) agreed that Inyo County would operate the Owens Valley town water systems with subsidies from the LADWP; (e) continued cooperative studies on the region's hydrology; and (f) inaugurated certain enhancement and mitigation projects.

The court approved the Interim Agreement and ordered the two parties to submit a groundwater management plan along with a new EIR prepared by the LADWP by 1989. This led to "A Preliminary Agreement Between the County of Inyo and the City of Los Angeles and Its Department of Water and Power on a Long Term Groundwater Management Plan for the Owens Valley and Inyo County," which was concluded on March 31, 1989, as well as a revised EIR.[29] Nevertheless, lingering disagreements over the details of the agreement and the EIR ensued, and both implementation of the long-term agreement and adoption of the EIR were delayed, first to 1991, and then until later.

The "Final Inyo/Los Angeles Long Term Water Agreement and Memorandum of Understanding (MOU)" went into effect on June 1997. It included a groundwater management plan funded largely by Los Angeles that would avoid causing significant changes in vegetation and environmental conditions. Further, all new wells were to be approved by both parties before installation, and the LADWP agreed to limit subsurface withdrawals in the Bishop cone. Los Angeles agreed to sell more of its properties in and around Bishop; to upgrade water systems in the towns and provide the water without charge; to allocate funds annually for saltcedar control, park rehabilitation, and maintenance; and to provide general financial assistance to Inyo County ($1,795,637 in 2002, for example).[30]

All in all, there were at least eleven major lawsuits brought against Los Angeles over its water export policies in Owens Valley between 1972 and 1997. The petitioners included (in various combinations) Inyo County, the Owens Valley Committee (a local lobbying group formed in 1983), the Sierra Club, the California Department of Fish and Game, and the State Lands Commission. With so many groups having potential standing in the litigation, settlement costs were increased, resulting in the seemingly unending continuation of legal action. The demands of the most extreme group could potentially cause any tentative settlement agreement and cooperation to unravel. This problem is even more severe under the public trust doctrine, which is examined in the following chapters.

THE LOWER OWENS RIVER PROJECT

In 1974 the State Water Resources Control Board (SWRCB) issued a license to the LADWP to appropriate and divert additional water from the Owens River for the second aqueduct. The license did not include requirements for a water bypass to protect fish stocks as outlined in the State Fish and Game Code, Sections 5937 and 5949. The SWRCB concluded that the city's Owens Valley operations predated the regulations.

But in *California Trout, Inc. v. State Water Resources Control Board* 207 Cal. App. 3d 584 (1989) and *California Trout, Inc. v. Superior Court* 218 Cal. App. 3d (1990), the so-called Cal Trout I and II cases, the court required that the diversion license be amended to require bypass flows. On May 29, 1991, the State Water Resources Control Board did so in Order 91-04, which required that the LADWP consult with the staff of SWRCB, the Department of Fish and Game, and the California Regional Water Quality Control Board to determine appropriate stream flows in Mono County. But the ruling also led to aqueduct releases for the Lower Owens River and its delta.

One element of the 1989 Groundwater Management Plan between Inyo County and Los Angeles already involved a proposal for releasing more aqueduct water to the Owens River as part of a proposed Lower Owens River Project (LORP). Under this project, a sixty-mile stretch of the Owens River below the aqueduct intake would be rewatered from the aqueduct to achieve a base flow of 40 cubic feet per second (cfs) for aquatic, riparian, and wetland habitats. To capture the water and return it to the aqueduct at the end of the new flow, the LADWP could construct a pumping station with a capacity of 50 cfs. Some of the water from the LORP, however, was to bypass the pump station to provide water to the Owens River delta for further habitat enhancement. The capital costs were estimated to be $7.5 million, at least half of which was to be paid by the LADWP along with annual operating expenses for the pumpback station.[31]

Following the 1991 SWRCB ruling, an Environmental Impact Report was to be prepared by the LADWP for the proposed Lower Owens River Project. Construction was to commence within three years of court approval of the EIR. In addition, the project was to undergo a California Environmental Quality Act review and have an Environmental Impact Statement (EIS) prepared as mandated by the National Environmental Policy Act (NEPA) for projects using federal funding. As a result, the environmental assessment preceding the LORP became a joint EIR/EIS prepared by the LADWP and approved by the Environmental Protection Agency.

Negotiations between various constituent groups, state agencies, Inyo County, and the LADWP over the LORP, along with studies of the environmental impact of rewatering sections of the river, followed between 1991 and 1997. In June 1997 a Memorandum of Understanding between Inyo County, the Owens Valley Committee, the Sierra Club, the California Department of Fish and Game, the State Lands Commission, and the LADWP was agreed to. It outlined habitat projects and provided for the formation of a Steering Committee with a representative from each party to the MOU to oversee their implementation. Restoration of water flows on the lower Owens River was to begin by June 2003.

A Draft EIR/EIS for the Lower Owens Valley Project was submitted by the LADWP, the EPA, and Inyo County in November 2002. The draft described options for the LORP and provided for a larger pumpback station of 150 cfs to return Owens River water to the aqueduct.[32] This draft environmental review, however, was not accepted by the court. The size of the pumpback station was one of the issues of contention because it would allow the LADWP not only to return water from the Lower Owens River Project, but also to access groundwater from the east side of lower Owens Valley, which had not been tapped previously. The plaintiffs were seeking to limit groundwater extraction by the LADWP, not to enhance it in other areas.

On September 26, 2003, the Sierra Club and the Owens Valley Committee sued in *Sierra Club et al. v. City of Los Angeles* (Sierra Club I) to enforce provisions of the MOU, including the overdue release of water in the river, and preparation of a revised EIR.[33] On December 4, 2003, California Attorney General Bill Lockyer also filed a lawsuit against Los Angeles to require restoration of the Lower Owens River. Later, in 2004, the LADWP, Inyo County, the Sierra Club, the Owens Valley Committee, the Department of Fish and Game, and the State Lands Commission settled the Sierra Club I MOU litigation through a Stipulation Order that included provisions for a 50 cfs pump station, final completion of the EIR by June 23, 2004, and release of water in the river by the end of 2004, as well as other payments to Inyo County and habitat mitigation investments from the LADWP.

Because of disagreements among the parties to the Stipulation Order, the LADWP completed the Final Environmental Impact Report independently. The document was issued in June 2004 with a 50 cfs pumpback station and other revisions to the Draft EIR/EIS.[34] The Final EIR was challenged again by the Sierra Club (Sierra Club II) in a series of court cases beginning in October 2004, and it was not approved by the Environmental Protection Agency, which withheld the federal funding for part of the LORP. In July 2005, the Inyo County Superior Court ordered water diversions to the lower Owens River from the aqueduct, as well as restrictions on LADWP pumping to replenish groundwater tables in order to provide the river with 40 cfs of water by January 2007.[35] If the deadline were not met, exports through the second aqueduct would be halted by the court.

In the Final EIR/EIS of 2004, the LADWP estimated the costs of the LORP as follows: upfront capital investments for the pumpback station, fencing, culverts, and similar items were $15.5 million; the discounted present value of annual operating and maintenance costs over fifteen years, not including the pump station costs, were $4.2 million; the present value of annual monitoring costs were $2.6 million; and the present value of mitigation program costs were $6.6 million. The total outlay was $28.9 million.[36] Many, if not most, of these costs were to be borne by the LADWP.[37] These figures do not include the sizable costs of the years of litigation and study as part of the conflict over the Lower Owens River Project.

There also were the opportunity costs of the lost water that was no longer available for export to Los Angeles via the aqueduct. The exact amount is unclear because of the complicated nature of the project. The mean 40 cfs of water diverted from the aqueduct to the river represents about 30,000 acre-feet of water annually, with some releases as high as 200 cfs during seasonal habitat flows.[38] The pump station was to capture up to 50 cfs and send the water to the Owens Lake Dust Mitigation Project, described below. Accordingly, this portion of the lost water would be accounted for in the dust control releases. At the same time, however, there would be water lost due to evapotranspiration from riparian vegetation, seepage into deep alluvial aquifers, and diversions to maintain off-river lakes, ponds, and habitat areas. Some water also was to be bypassed on to the Owens River delta.[39]

The initial water losses were projected to be 66,818 acre-feet until subsurface aquifers filled, and then 36,958 acre-feet annually as a steady state.[40] Taking $501 per acre-foot as the replacement cost of water (the mean price of tier 1 and tier 2 Metropolitan Water District water), this steady-state loss translates to $18,516,000 annually with a present value of $221,042,806 using the same time period and interest rate as in the EIR.[41] By this account, the annual replacement water, capital, and operating costs of the Lower Owens River Project are $47.4 million. No cost-benefit analysis was conducted.

LOS ANGELES'S WATER RIGHTS AND
OWENS LAKE DUST MITIGATION

Closely related to the Lower Owens River Project was the Owens Lake Dust Mitigation Program, overseen by the Great Basin Unified Air Pollution Control

District (GBUAPCD). Although the level of Owens Lake had fluctuated in the late nineteenth and early twentieth centuries owing to annual runoff variation and agricultural diversions in the northern Owens Valley, the Los Angeles Aqueduct effectively intercepted the residual water. The lake dried up around 1923. This left over 60 square miles of dry lakebed exposed to spring and fall winds. The soft salt crust was pulverized by saltating sand. Tiny airborne particles with a diameter one-tenth the width of a human hair (equal to 10 microns or PM-10) were carried aloft in dust storms.

Owens Lake was cited as the largest source of fugitive dust pollution in North America in the early 1990s.[42] Some 290,000 tons of PM-10 per year were estimated to be emitted, exceeding National Ambient Air Quality Standards for particulate matter in the immediate area, as well as in the neighboring small communities of Lone Pine and Keeler in Owens Valley. Dust pollution was also a nuisance and potentially a health hazard in the more populous Indian Wells Valley.[43] The particles were small enough to be inhaled deeply into the lower respiratory tract, causing breathing difficulties and asthma attacks, as well as eye and nasal irritation. Other potentially harmful effects included damage to sensitive ecosystems, such as those of the nearby bristlecone pine, and curtailment of the flight operations at the China Lake Naval Weapons Center as often as twelve days a year.[44]

In 1998 the Great Basin Unified Air Pollution Control District adopted a State Implementation Plan (SIP), revised in 2003, to address the problem, and it identified the dust control measures to be implemented by the LADWP on the Owens Lakebed.[45] These included shallow flooding, planting vegetated areas in salt grass, and providing gravel blankets four inches thick to reduce dust emissions over 35 square miles of lake. A Memorandum of Agreement (MOA) between the GBUAPCD and the LADWP was approved by the Environmental Protection Agency in October 1999. The MOA required that the LADWP implement the approved dust abatement measures to meet federal air-quality standards by 2006. Pollution control costs were to be paid by Los Angeles.[46]

In response, in November and December 2001 the LADWP began shallow flooding of 11.9 square miles of the lakebed. The 2004 EIR/EIS for the Lower Owens River Project forecast annual water requirements for dust mitigation on the Owens Lakebed of 51,000 acre-feet initially, rising to 64,700 acre-feet by 2006.[47] The 2003 State Implementation Plan adopted by the Great Basin Unified Air Pollution Control District estimated the capital costs for Owens Lake dust control to be $415 million. This figure included the costs of building berms, dams, drainage systems, fences, and roadways; installing pumping equipment for moving and recirculating water, as well as for blending fresh with highly saline water; putting in drip systems with emitters for vegetation; and spreading gravel. These were to cover some 29.8 square miles of lakebed.[48] Converting this capital investment figure to an annual flow for fifteen years at 3 percent (the same parameters used above) results in an annual cost of approximately $35 million per year. The operations and maintenance costs for the LORP were projected to be $27 million annually. These annual costs total to $62 million for dust control on the Owens Lakebed.[49] The present value of these annual expenses is $740,151,975 for fifteen years at 3 percent.

The operations expenses are said to include the annual cost of water replacement from the Metropolitan Water District. But the figure may be too low. One estimate of the opportunity cost of the water is nearly $21 million, and if that is accurate, operations and maintenance could be reflected in the remaining $6 million annually ($27–21 million).[50] The operations on Owens Lake are subject to clogging and silting and require substantial labor input, so the maintenance costs are apt to be considerable. If, however, 55,000 acre-feet is used annually as outlined in the 2005 LADWP Urban Water Management Plan, then with a replacement cost of $501 per acre-foot, the opportunity cost of the water alone is $27.6 million, with no budgeting for operations and maintenance.[51] The present value of this water over fifteen years at 3 percent would be $329,487,000. Accordingly, the overall investment in capital and annual operations involves considerable resources for PM-10 control on the Owens Lakebed.

Because the Owens Lake Dust Mitigation Program was initiated under the Clean Air Act, no cost-benefit analysis was required. While the costs of pollution abatement are fairly easy to quantify and appear to be large, borne virtually entirely by the ratepayers of Los Angeles, the benefit calculations are less straightforward. Despite repeated references to the damaging health effects of Owens Lake PM-10 dust, there apparently are no formal epidemiological studies that document or measure its magnitude in the affected region. The references rely on anecdotal reports of respiratory irritations and illnesses in nearby locales.[52]

Comparisons of the prevalence of lung cancer, pediatric and adult asthma, chronic bronchitis, and emphysema in Inyo and Mono Counties with the incidence of those diseases in the state of California as a whole, as well as similar comparisons of asthma hospitalization and mortality rates, however, do not indicate greater health risks in those counties.[53] There also are apparently no published studies of the impact of the pollution on animals or plants.[54] Nor does there seem to be systematic data on the alleged effects of pollution on trees and other plants in nearby wilderness areas. And the impact on operations on the

China Lake Naval Air Weapons Station remains inconclusive, although it might have been important. As a result, there is little definitive quantitative evidence of the nature or magnitude of the economic benefits of mitigating PM-10 dust pollution off the dry Owens Lakebed. This is despite the sizable investment mandated for dust pollution control.

CONCLUSION

This chapter has detailed the lingering conflict over the water rights acquired by the LADWP in the 1920s. The agency managed its land holdings in Owens Valley to supply water. Agricultural, recreational, and other amenity uses were residual demands, serviced only when there was excess water beyond what was needed for the aqueduct. Over time, however, new values and new claims for Los Angeles's water emerged just as the city was planning to transmit more water through the second aqueduct. Demands for reallocation of water to maintain the region's rural agricultural and recreational economy and to address physical externalities associated with groundwater pumping and with water export out of the region were addressed through the courts. This course of action was extremely controversial, with a seemingly endless series of litigation and regulatory actions. A similar slow and litigious process is described for the Mono Basin in the following chapter.

7

WATER RIGHTS AND WATER REALLOCATION IN THE MONO BASIN, 1935–2006

> Mono Lake is one cog in the water system feeding Los Angeles. It is also one of the oldest continuously existing lakes in North America. Nestled in the high Sierras, Mono Lake is also a critical wildlife habitat. And it is caught in a battle for its survival.
>
> *Peter Goin, Review of* At Mono Lake, *1984*

By the early 1930s, Los Angeles owned virtually all the land and water rights in Owens Valley, and the city was acquiring additional land and water rights in the adjacent Mono Basin to the north. All of this activity was to supply high-quality, gravity-flow water to the Los Angeles Aqueduct and thereby to the booming population and commercial growth in the city that depended upon it. Through its acquisitions, the Los Angeles Department of Water and Power (LADWP) redirected Owens Valley water from marginal farming to urban use. It sought to do the same in Mono County, where conditions for agriculture were even less favorable.

Although the Mono Basin had a lot of water, it had fewer people than Owens Valley. It was higher in elevation, had less arable land, and was more isolated. Some water was used to irrigate pastures, but most flowed in streams to Mono Lake, a landlocked, highly alkaline body of water, somewhat similar to, although deeper than, Owens Lake to the south.

To gain Mono County water, the LADWP also bought the land and water rights in the region and extended the Los Angeles Aqueduct north through the Mono Craters Tunnel. While this was occurring, other water was being brought to southern California by the construction of Hoover Dam and the Colorado River Aqueduct. Those waters would arrive by 1941 and be used largely by other communities. Los Angeles had higher-quality and lower-cost water from Owens Valley and the Mono Basin.

Because water transmission and use requires fixed, nondeployable capital investments in aqueducts, canals, dams, and other infrastructure, Los Angeles also was tied to Owens Valley and the Mono Basin. Its urban population and economic growth depended on access to secure, reliable supplies of high-quality water from those regions. The LADWP was mandated to provide it. For all of these reasons, the agency fought tenaciously forty years later to retain control over the water as new demands for it arose.

ACQUISITION OF WATER RIGHTS BY LOS ANGELES

In the late 1920s, as the LADWP was nearing the completion of its acquisition of the water-bearing lands in Owens Valley, it began increasing its effort to locate alternative sources of supply for the aqueduct. In 1930, Los Angeles's population passed 1.2 million people, up from 577,000 in 1920, an increase of 115 percent. A natural source of more water was the Mono Basin just to the north with similar high-quality water that could be directed by gravity flow to the Los Angeles Aqueduct. Accessing the water would require construction of a tunnel from the basin to the Owens River drainage and diversion of flows from four tributary streams that fed Mono Lake: Rush, Lee Vining, Walker, and Parker Creeks. Another attraction of the move north was that the additional water could be directed through the Owens River Gorge to generate some 268 million kilowatt hours (kWhs) of electricity per year. By contrast, pumping Colorado River water to southern California was expected to *require* 186 million kWhs per year.[1]

Armed with funds from the May 20, 1930, bond issue, land agents from the Water Board began to acquire properties and water rights above and around Mono Lake. Of the $38 million authorized by the bonds, $7 million was for purchase of land in the Mono Basin (principally properties of the Southern Sierras Power Company and Cain Irrigation Company); $1.45 million for construction of a conduit to carry water from Lee Vining Creek to Silver Lake; $550,000 for construction of Silver Lake Dam; $5.5 million to build an 11.3-mile tunnel from Silver Lake to the Owens River via Long Valley; $750,000 for a dam and storage reservoir in Long Valley; $6.66 million to purchase remaining private farmland in Long Valley and Owens Valley; and $600,000 to raise the capacity of the aqueduct from 400 cfs to at least 440 cfs to carry the new water flow.[2] Not all of the bonds were immediately issued because of the deterioration in the bond market with the advent of the Great Depression. The city sought Reconstruction Finance Corporation purchase of $20.8 million of the bonds and a grant of $7 million.[3]

Unlike in Owens Valley, where much of the water was claimed under the appropriative rights doctrine and channeled to farms and ranches via ditches, in the Mono Basin most of the water was claimed under the riparian rights doctrine to irrigate property adjacent to the four streams and to generate electrical power via small hydroelectric facilities. As with Owens Valley agriculture, irrigation in the Mono Basin involved extravagant application of water, with the largest operation, the Cain Irrigation Company, using approximately 6 acre-feet of water per acre of land and in some cases as much as 43 acre-feet per acre for very low-value alfalfa and pasture.[4] At the time, there was little alternative use for the water, and much of it percolated back to the streams and ultimately to Mono Lake.

In Owens Valley the LADWP had not used condemnation via eminent domain to secure land and water rights, but in the Mono Basin it did so.[5] The city faced a much more severe holdout problem in the Mono case since it had to buy all riparian claims on a stream before it could divert any water. Each riparian holder had the right to an undiminished stream flow, a condition that would block any diversion out of the streambed to the Los Angeles Aqueduct.

The vulnerability to holdout was stressed by City Attorney Erwin Werner in a December 8, 1930, letter to the Board of Water and Power Commissioners. He argued that any riparian water rights acquired by the city would not include a right to change place of use and diversion of water to Owens Valley if *any* other water rights holder on the stream could show a present or threatened injury from such action. Accordingly, he urged that the city acquire *all* water rights on each stream either by purchase or by condemnation in order to export water.[6]

As a result, on March 17, 1931, the city filed condemnation proceedings as part of the case *Los Angeles v. Nina B. Aiken, et al.*[7] The complaint named 130 defendants as having water rights in the Mono Basin that the city desired. As with Owens Valley lands, there were problems with valuation and the determination of compensation. During the winter of 1933–34, Clarence Hill, a Water Board right-of-way and land agent, hired special appraisers to investigate the properties and to submit valuation reports for use in the *Aiken* trial. On Lee Vining Creek, valuation was fairly simple because the Cain Irrigation Company held all the riparian rights, but things were more complicated on Rush Creek, where there were more rights holders.[8] The suit led some of the property owners to lower the prices they were demanding. For example, on July 31, 1928, representatives of Southern Sierras Power had offered to sell most of their properties for $8.15 million, but after the *Aiken* lawsuit was initiated, they lowered their asking price to $6.95 million.[9]

The condemnation case continued through 1935, when it was appealed to the California Supreme Court in *City of Los Angeles et al. v. Aiken et al.* (10 Cal. App. 2d 460). Eventually the disputes over compensation were resolved and the targeted properties were purchased. By 1938 the Water Board had acquired 16,340 acres of private land and the water rights on another 9,008 acres in the Mono

Basin. Ultimately, over 30,000 acres were purchased in the region during the 1930s.[10] There were also substantial withdrawals of federal lands from private entry around Mono Lake to protect Los Angeles's water supplies and to provide for right-of-way for the aqueduct extension. Some of those lands also were later transferred or sold to Los Angeles.[11]

With the move to Mono, the Los Angeles Aqueduct was extended another 105 miles. Its northern end at the Lee Vining intake was 338 miles from Los Angeles, effectively due east of the Golden Gate Bridge. The project included Grant Lake Dam on Rush Creek, the Lee Vining conduit to the Grant Lake Reservoir, and the Mono Craters Tunnel to the Owens River. After five years of construction, employing 1,800 workers and an expenditure of $40 million, the venture was completed in April 1941.[12] Figure 7.1 illustrates the Los Angeles Aqueduct system with the Mono Basin extension.[13]

WATER EXPORTS FROM THE MONO BASIN TO LOS ANGELES

In 1940, the LADWP was granted permits by the Division of Water Resources (later the State Water Rights Board and the State Water Resources Control Board, SWRCB) to appropriate the complete flows from Rush, Lee Vining, Parker, and Walker Creeks for its new Mono Basin export system.[14] The limited downstream capacity of the Los Angeles Aqueduct, however, prevented full appropriation of the water until the second aqueduct was completed in 1970.

The failure to fully appropriate the stream flows for the aqueduct subsequently became a vehicle for weakening Los Angeles's legal claim to Mono water. Under the California Constitution (article X, section 2 amendment) all appropriated waters of the state were to be put to reasonable and beneficial use, with any surplus available for claiming by others.[15]

Although sensible at an earlier time of initial water allocation during the Gold Rush to prevent speculative appropriation in a semi-arid region, continued adherence to the beneficial use requirement for appropriative water rights placed the security of any water reserved for future access into question.[16] So long as there were no competing claimants, the lack of immediate beneficial use was not a problem for Los Angeles. But once this situation changed, there would be trouble.

The first regular flow of water from the Mono Basin to the Owens River above the Long Valley Dam and the Owens River Gorge began in 1941.[17] Earlier in 1935, when the LADWP had applied for a license to construct Grant Lake Dam and to store water behind it without a fish ladder or water release, the State Fish and

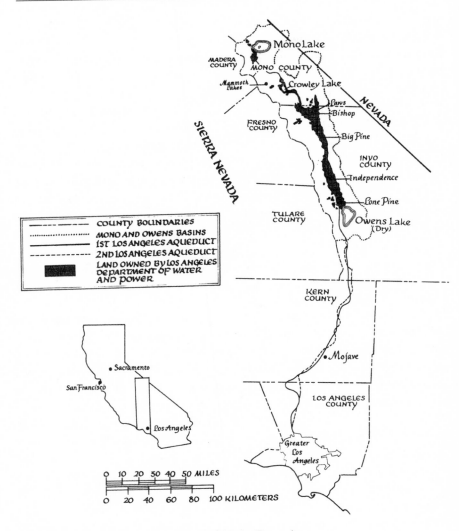

Figure 7.1 Los Angeles Aqueduct and Mono Extension.

SOURCE: Reprinted from Hundley (2001, 159).

Game Department approved the application. But in 1936, when the Long Valley Dam was being constructed, the department called for protection of fish habitat. Section 5937 of the Fish and Game Code required that the owner of a dam provide sufficient water at all times to keep fish in good condition below the dam.

In lieu of this specific requirement, on September 12, 1940, the LADWP reached an agreement with the Department of Fish and Game to provide a fish hatchery on Hot Creek to augment fish stocks in area streams and reservoirs where water remained.[18] This 1940 Hot Creek Agreement allowed for the drying up of the Mono creeks below their diversion dams, and eventually sixteen miles of the Owens River below the Long Valley Dam as well. The Hot Creek Agreement, however, was later to be rejected as providing insufficient compliance with the State Fish and Game Code.

Concern about fish habitat in the upper Owens River rose in 1952 as water was being diverted from the Long Valley Dam to be used by the new hydroelectric plants just completed in the Owens River Gorge. This action left the river between the Upper and Center Gorge power plants without water, and fishing groups lobbied for legislation to require delivery of enough water to sustain the fishery. In response, the legislature enacted Fish and Game Code Section 5946 in 1953, which stated that preliminary permits or final licenses for water diversion in Inyo and Mono Counties were conditional on a water release according to Section 5937 for protection of fish environments.

The Fish and Game Department applied the new code to Los Angeles's diversion request for the hydroelectric sites, but the attorney general ruled that the city's request would be governed by the earlier 1940 Hot Creek Agreement. Later in 1955, when the city applied for permanent diversion licenses, the attorney general also granted them the same immunity as the 1940 permits. This ruling was unchallenged for thirty years, but then it would be overturned.[19]

INCREASED WATER DIVERSIONS AND LEGAL CONFLICT

By the early 1960s it was time for Los Angeles to draw on more of the city's water in the Mono Basin. The Los Angeles Aqueduct was at full capacity, and urban demand continued to grow: Los Angeles's population reached 2,479,000 people in 1960. There also was increased anxiety over the status of the city's claim to the water it had not used under its 1940 diversion permit.[20] In 1956, the State Department of Water Resources (DWR) reported that Los Angeles was exporting only 320,000 acre-feet of the 590,000 acre-feet annually available in Owens Valley and the Mono Basin. In 1959 the State Water Rights Board warned that Los Angeles could lose its rights to the water it was not appropriating, noting growing interest in the apparently surplus water.[21] Further, the legislative representatives of Inyo and Mono Counties sought studies of how the excess water might be used in Owens Valley and the Mono Basin.

In July 1963, construction began on the second aqueduct, and it was completed in 1970.[22] The second aqueduct could be filled with water from increased

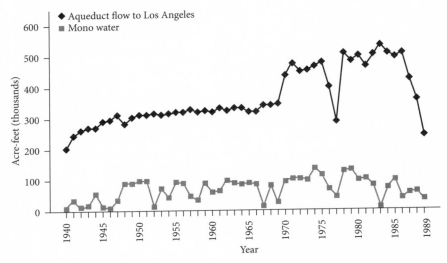

Figure 7.2 Los Angeles Aqueduct and Mono Basin Flows, 1940–89.

SOURCE: Data for the figure from Jones and Stokes (1993, 3A-6), and for Mono flows, personal communication with Paul Scantlin, Civil Engineer, Water Department, LADWP. These include monthly water exports through Mono Craters Tunnel. Monthly data were summed for January–December for each year.

NOTE: Aqueduct flow totals vary somewhat across sources, most likely due to differences in months included and in conversion factors.

groundwater pumping and reduced irrigation in Owens Valley and from greater diversions from Mono. To gain authorization for the additional export of water from the Mono Basin, the LADWP applied to the State Water Resources Control Board for permanent diversion licenses. In 1974, diversion licenses 10191 and 10192 were granted, allowing the city to divert up to 167,000 acre-feet annually. This figure was never reached, however.

Figure 7.2 illustrates the total Los Angeles Aqueduct flow and the contribution from the Mono Basin from 1940 through 1989. Between 1940 and 1970 an average of 57,067 acre-feet was exported from the region, although the amounts fluctuated considerably.[23] The Mono Basin seems to have been primarily a residual source of water. At their peak, Mono exports provided about 15 percent of Los Angeles's water supply. Additional power also was generated through the Owens River Gorge.[24]

With the new capacity afforded by the second aqueduct, exports jumped from about 21,000 acre-feet in 1969 to close to 100,000 acre-feet or more through 1975. The peak was nearly 135,000 acre-feet in 1974. With new court-ordered restrictions, exports dropped to 41,000 acre-feet in 1977. They rebounded with the relaxation of restrictions during drought in the late 1970s, but then fell to near zero in 1983, rising and fluctuating again during the course of litigation. In 1989, approximately 31,000 acre-feet were shipped.[25]

With larger interception of tributary flows, Rush, Lee Vining, Parker, and Walker Creeks dried up below the diversion points, and the level of Mono Lake began to decline about 1.6 feet a year.[26] Between 1941 and 1981 the lake's level had fallen about 46 feet, with one-third of that decline occurring after 1970. The surface area of Mono Lake receded from 90 to 60 square miles, and its salinity increased from 50 grams per liter to 90 grams per liter.[27] The resulting effects, first on the lake and then on stream fish habitats, brought growing opposition to the water diversions.

Concern rose about the impact of the lower lake levels and greater salinity on aquatic wildlife (brine shrimp and flies) and the migrating bird population, particularly the California Gulls that nested around and on islands in the lake.[28] There were fears about the effects of any decline in the gulls' Mono Lake food sources and the increased vulnerability of their nests to predators as lake levels retreated. Larger dust storms off the exposed dry lakebed raised complaints similar to those starting to be voiced about dust pollution from the dry Owens Lakebed.[29]

By the late 1970s, there were bills in the California legislature to halt the water diversions; environmental groups were rallying to generate political support for government intervention; court suits were filed; and petitions were drafted to save Mono Lake.[30] In 1978 the Resources Agency of California assembled a special task force to examine the condition of Mono Lake, and in 1979 it called for a dramatic cut of 85 percent in Mono water exports, from 100,000 to 15,000 acre-feet annually. The agency argued that this move would allow the lake's level to rise to its 1970 level of 6,388 feet. The task force estimated that its recommendations would cost $250 million in lost water and power, but it predicted that 80 percent could be recovered by conservation in Los Angeles and wastewater reclamation. The costs of alternative water supplies were to be split among the state of California (36 percent), Los Angeles (36 percent), and the federal government (28 percent). The costs of wastewater reclamation and conservation, as well as those resulting from lost power generation, were to be borne by Los Angeles alone. The LADWP countered that the outlays would be closer to $2 billion and claimed that it was unfair to have the city bear so much of the cost of restoring the lake's level.[31]

THE PUBLIC TRUST DOCTRINE RULINGS

The National Audubon Society, Friends of the Earth, the Sierra Club, and a new coalition of environmental activists, the Mono Lake Committee that had formed in 1978, brought suit in 1979 to curtail export of water under the public trust doctrine. Referring to *Marks v. Whitney*, 6 Cal. 3d 251 (1971), which held that the public trust doctrine applied not only to navigable waterways but also to streams used for recreation, wildlife habitat, and ecological study, the plaintiffs charged that Mono Lake was being harmed and that the diversion was not a reasonable and beneficial use as required by the state's appropriative water rights system.[32]

On November 9, 1981, the Alpine County Superior Court rejected this public trust challenge to Los Angeles's water rights. The judge argued that administrative remedies should have been exhausted prior to resort to litigation and that the public trust doctrine was subsumed in existing California water rights law.[33]

The plaintiffs' appeal to the California Supreme Court was successful, however, and the ruling set precedent not only for addressing the Mono water rights controversy, but also for extending the public trust doctrine more broadly to environmental regulation in other contexts and in other states.[34] On February 17, 1983, in *National Audubon Society v. Superior Court*, the court held that exercise of appropriative water rights is subject to limitation by the state in order to protect public trust values, including those of wildlife habitat: "Thus, the public trust is more than an affirmation of state power to use public property for public purposes. It is an affirmation of the duty of the state to protect the people's common heritage of streams, lakes, marshlands and tidelands" (33 Cal 3d 441).

According to the court, public trust regulatory responsibilities applied ex post to existing water rights, and these rights were use rights only that could be reconsidered in light of changing perceptions of the trust. Water belonged to the people. The court charged the State Water Resources Control Board with monitoring water use and reallocating it in a manner consistent with the public trust: "Thus, the function of the Water Board has steadily evolved from the narrow role of deciding priorities between competing appropriators to the charge of comprehensive planning and allocation of waters" (33 Cal 3d 444).

The court held that Los Angeles's water rights in Owens Valley and the Mono Basin were provisional and subject to periodic adjustment by the legislature and its authorized agencies. The effect of such a broad interpretation of the public trust doctrine on property rights and incentives for new investment in water and water-related infrastructure was controversial at the time. They are discussed in the next chapter.

Because the ruling signaled not only the mostly uncompensated loss of valuable water rights, but also the value of Los Angeles's past fixed investments in the aqueducts, dams, reservoirs, and hydroelectric facilities, the LADWP filed a petition for certiorari with the United States Supreme Court on two grounds: first, that the public trust doctrine was a federal doctrine and the California court misinterpreted it; second, that the decision deprived Los Angeles of vested property rights without due process of law (a takings). The Department of the Interior's Regional Solicitor for California also agreed that the California State Supreme Court's ruling fundamentally changed the nature of Los Angeles's water rights.[35] The appeal, however, was denied on November 7, 1983.[36]

In July 1983, the U.S. District Circuit Court in Sacramento ordered the LADWP to reduce its water diversions from the Mono Basin through August 1984 in order to release enough water to stabilize the lake's level.[37] In August 1984 the city's diversions again resumed from all but Rush Creek, and export volumes increased, as shown in Figure 7.2. Because there were trout stranded in the stream, the Department of Fish and Game, along with an advocacy group, California Trout, argued that the city should allow the flows in Rush Creek, the Mono Basin's largest stream, to continue.

California Trout joined with the National Audubon Society, the Mono Lake Committee, and others in suing for continuation of the Rush Creek releases under the public trust doctrine.[38] On March 7, 1985, in *Dahlgren v. Los Angeles* (Mono County Superior Court No. 8092), the so-called Rush Creek Case, the court issued a restraining order requiring a flow of 19 cfs to provide fish habitat as part of the public trust. Later, in August 1985, the court extended the order while studies were conducted to determine the amount of water necessary to maintain fish habitat. These studies took six years to complete.[39] After 1985, Mono flows to the aqueduct declined, as indicated in Figure 7.2. A similar court ruling in 1987 in *Mono Lake Committee v. Los Angeles* (Mono County Superior Court No. 8608), the so-called Lee Vining Creek Case, required the resumption of water flows of 4 to 5 cfs down Lee Vining Creek to protect public trust values.

More permanent revision of Los Angeles's Mono water rights occurred in *California Trout v. State Water Resources Control Board* (207 Cal. App. 3d 585) on January 26, 1989 (Cal Trout I), where the appeals court concluded that Los Angeles's 1974 diversion licenses should be revoked and reissued because they did not comply with Fish and Game Code Section 5946, which required sufficient water to be released to protect fish habitat. The court overturned an earlier opinion by the Sacramento County Superior Court issued on July 30, 1984, that the city's appropriative rights were immune from such a challenge according to the terms of the 1940 Hot Creek Agreement. The appeals court held that since Los Angeles's 1940 diversion permits had not been placed fully into beneficial use until 1974, they were subject to the requirements of Section 5946.

In August 1987, a National Academy of Sciences report, *The Mono Basin Ecosystem: Effects of Changing Lake Level*, suggested that a lake level of 6,380 feet be maintained to protect the lake's ecosystem. Another state-funded study called for a similar minimum level.[40]

Two years later, on August 22, 1989, in *The Matter of Mono Lake Water Rights Cases* (El Dorado County Superior Court Coordinated Proceedings Nos. 2284, 2288), the court issued an injunction halting export of all water from the Mono Basin through March 30, 1990, and releases of 85 to 100 cfs down Rush Creek and 60 cfs down Lee Vining Creek to stabilize the lake's level at 6,377 feet above sea level.

To provide some financial reimbursement to Los Angeles for the lost Mono water, the California legislature passed AB444, the Environmental Water Act of 1989, on September 22, 1989, which allocated $60 million for alternative water sources. Funds would be granted, however, only upon joint application by the LADWP and the Mono Lake Committee, a requirement that gave equal standing to one of Los Angeles's key competitors for its water.[41] They would have to jointly agree on alternatives, and doing so would imply that the city's past water rights were no longer in effect. A Draft Environmental Impact Report to examine the effects of water export from the basin and to outline management options also was to be prepared by the State Water Resources Control Board.

On February 23, 1990, in *California Trout v. Superior Court*, 218 Cal. App. 3d 187 (Cal Trout II), the Third District Court of Appeal further directed that the SWRCB amend the LADWP's 1974 diversion licenses to include the requirement that: "The licensee shall release sufficient water into the streams from its dams to reestablish and maintain the fisheries which existed in them prior to its diversion of water." This ruling mandated that Rush, Lee Vining, Parker, and Walker Creeks be restored to their 1940 status. Undiverted stream flows were made the baseline for sustaining fish habitat. No cost-benefit study was required. Some 60,000 acre-feet per year were to be released to the streams and Mono Lake.[42] A Restoration Technical Committee with one seat each for the LADWP, the Mono Lake Committee, the National Audubon Society, California Trout, and the Department of Fish and Game, was to manage the restoration of aquatic and riparian habitats. The LADWP was to pay for the restoration.[43]

Disputes among these groups over the appropriate lake level target, the amount of water diversions to be allowed, and the extent of habitat restoration, however, brought another round of litigation. On April 17, 1991, the El Dorado County Superior Court ordered that the lake level be held at 6,377 feet and required the LADWP to pay court costs.

There still was no agreement between the LADWP and the Mono Lake Committee on the allocation of the funds set aside by the state under AB444. In September 1992, the U.S. Congress passed HR429, the Reclamation Projects Authorization and Adjustment Act, authorizing the Bureau of Reclamation to pay one-fourth of the cost of some water recycling projects, conservation, and effluent recycling in southern California as some offset for lost Mono water.[44]

Further pressure was added to the LADWP to give up more Mono water when the Environmental Protection Agency ruled on July 7, 1993, that the Mono Basin was in moderate nonattainment of Federal Clean Air Act standards due to blowing dust from the dry Mono Lakebed. Regulatory jurisdiction for meeting federal standards was placed under the Great Basin Unified Air Pollution Control District, the same agency that was monitoring investments by the LADWP to mitigate dust pollution from the Owens Lakebed.[45]

After nearly twenty years of conflict, the judicial reallocation of the city's water rights was becoming extremely costly. In 1991, the LADWP estimated that it had spent up to $12 million for outside lawyers and consultants since 1979.[46] In 1993, the department predicted that the long-term costs of replacing Mono water could reach $1 billion.[47] The legislature-mandated *Draft Environmental Impact Report for the Review of the Mono Basin Water Rights of the City of Los Angeles* that was released in May 1993 with twenty-eight technical reports cost the LADWP another $4.1 million.[48]

The Draft EIR presented a lake-level benchmark of 6,390 feet that would end dust pollution and maintain its remarkable tufa formations. The Great Basin Unified Air Pollution Control District, the State Department of Parks and Recreation, the State Lands Commission, and other agencies supported this benchmark or higher targets. Even so, the lake level necessary for sustaining wildlife habitats and for protecting fish stocks in the streams still was not fully known.

On September 16, 1994, the SWRCB published the Final Environmental Impact Report, which called for a target lake level of 6,390 feet. To achieve it, there could be no water diversions by the LADWP from the Mono Basin until the lake reached 6,377 feet; then 4,500 acre-feet a year could be withdrawn until the lake was at 6,390 feet; after that, 16,000 acre-feet could be exported until the lake was at 6,391; and at higher levels all water in excess of flows necessary to protect fish habitat could be diverted, for an average of 30,800 acre-feet per year. This process would take about twenty years. These final exports would be about one-third the amount diverted by the city in the early 1970s.[49]

Finally, on September 28, 1994, the State Water Resources Control Board formally amended the LADWP's Mono water rights through Decision D-1631. As ordered in 1983 in the *Audubon* case and in 1989 and 1990 in the Cal Trout I and Cal Trout II cases, the exports allowed by the diversion licenses issued twenty years earlier were reduced to comply with Fish and Game Code Sections 5937 and 5946 and to protect public trust values in Mono Lake. Only small diversions would be allowed until the lake level reached 6,392 feet above sea level.

In their 1993 Draft Environmental Impact Report, Jones and Stokes Associates estimated that the cost to the city of these restrictions on the export of water from the Mono Basin would be $35,460,000 annually owing to the opportunity cost of the lost water and power that could not be generated from it. This figure did not include the lost capital value of the infrastructure that would be under-utilized.[50] The present value of these annual costs over fifteen years at 3 percent is $423,319,178.

CONCLUSION

This chapter has described how Los Angeles acquired water rights in the Mono Basin in the 1930s and made investments in infrastructure to connect the basin to the Los Angeles Aqueduct in Owens Valley. Regional agriculture, recreation, wildlife, and the integrity of Mono Lake were subsidiary in importance to meeting growing urban water demand. Moreover the cumulative impact of withdrawal did not become apparent until much later. Because of a lack of export capacity, the LADWP did not fully exercise its diversion options until completion of the second aqueduct in 1970. Although there were substantial water shipments from the Mono Basin from time to time, amounts over 100,000 acre-feet were not shipped until after 1970.

With these increased diversions, however, concerns about the environmental impact of water export intensified. Environmental groups organized an effective political and public relations campaign to save Mono Lake, its tufa formations, and the habitat it provided for waterfowl. A complicated series of legal challenges to Los Angeles's water rights and diversions took place simultaneously. These were highlighted by the famous public trust ruling in *National Audubon Society v. Superior Court* in 1983. For Owens Valley, the litigation process was contentious, slow, and very costly. It may have delayed the response to the diversion externalities. Moreover, the notoriety of the Mono Lake situation and the nature of the judicial opinions rendered in regard to it potentially had very profound implications for the security of water rights and actions to be taken to respond to changing water values and externalities throughout the West.

8 | THE COSTS OF JUDICIAL REALLOCATION OF WATER RIGHTS AND THE PUBLIC TRUST DOCTRINE

> In Mono Lake, the California Supreme Court, in just a few pages, thoroughly modernized California water law. Because of its scarcity and its importance to so many other resource uses, water is the most important natural resource in the American West. Unfortunately, Western water use is largely governed by an archaic rule of temporal priority, the prior appropriation of water.
>
> *Michael Blumm and Thea Schwartz, "Mono Lake and the Evolving Public Trust in Western Water," 1995*

This chapter reviews the history of litigation underlying the revision of Los Angeles's water rights to illustrate how the course of action seems to have increased the transaction costs of addressing environmental and recreational concerns and delayed the response to them. The alternative of greater reliance on compensated exchanges, either through market purchases where possible, or through regulatory proceedings where not, is explored.

Since the most substantial challenge to appropriative water rights occurred though the public trust ruling in the *Mono Lake* case of 1983, the doctrine is examined in more detail as to its implications for settlement costs in disputes, the generation of information for judicial decisions regarding the division of water, and the security of any allocation.[1] There was great enthusiasm after 1983 among advocates of extending the public trust to other water and environmental conflicts.[2] But its application appears to have been more limited than anticipated at the time. The costs identified here may have led to more cautious extension of the doctrine.[3]

A SUMMARY OF REGULATORY ACTIONS TAKEN REGARDING LOS ANGELES WATER RIGHTS

Table 8.1 summarizes the judicial and administrative actions affecting Los Angeles's water rights, along with other milestones in the history of its water rights in Inyo and Mono Counties.

The entries in the table are impressive in their number, repetition, and scope in challenging Los Angeles's water rights in Owens Valley and the Mono Basin. Judicial and regulatory actions began in 1973 and were ongoing as of 2005,

TABLE 8.1
A chronology of the regulatory actions regarding Los Angeles's water rights

Year	Court case or agency	Action
1938		Los Angeles owns 99 percent of farmland and water rights and 88 percent of town property in Owens Valley. Aqueduct supplies over 70 percent of city water.
1940	State Water Rights Board	LADWP granted permits to appropriate Mono water for the aqueduct.
	State Fish and Game Commission	Hot Creek Agreement to satisfy Fish and Game Code Section 5937.
1953	State Fish and Game Commission	Fish and Game Code Section 5946 holds preliminary permits or final licenses for water diversion in Inyo and Mono Counties conditional on a water release under Section 5937.
1970		Los Angeles completes the second aqueduct.
	California legislature	California Environmental Quality Act (CEQA) enacted.
1973	*County of Inyo v. Yorty*, 32 Cal. App. 3d 795	CEQA applies to groundwater pumping for the aqueduct; EIR required; pumping held to half of capacity.
1974	State Water Resources Control Board (SWRCB)	LADWP receives permanent licenses 10191 and 10192 to divert up to 167,000 a.f. annually from Mono.
1977	*County of Inyo v. City of Los Angeles*, 71 Cal. App. 3d, 185	EIR insufficient; groundwater pumping and water allocations within Owens Valley remained regulated.
1980	Inyo County	Ordinance to regulate the extraction of groundwater within the Owens Valley groundwater basin.
1983	*National Audubon Society v. Superior Court*, 33 Cal 3d 419	Appropriative water rights restricted by the public trust doctrine.
1984	*County of Inyo v. City of Los Angeles*, 160 Cal. App.3d 1178	Court requires water plan by LADWP and Inyo County.
1985	*Dahlgren v. Los Angeles*, Mono County Superior Court No. 8092	Public trust requires release of 19 cfs down Rush Creek to provide fish habitat.
1987	Environmental Protection Agency	Owens Valley found to be in "nonattainment" of federal PM-10 standards for particulate matter.
	Mono Lake Committee v. Los Angeles, Mono County Superior Court No. 8608	Public trust requires release of 4–5 cfs down Lee Vining Creek.

TABLE 8.1

(Continued)

Year	Court case or agency	Action
1989	*California Trout v. State Water Resources Control Board*, 207 Cal. App. 3d 585	Los Angeles's diversion licenses revoked and reissued to comply with Fish and Game Code Sections 5946 and 5937.
	The Matter of Mono Lake Water Rights Cases, El Dorado County Superior Court Coordinated Proceedings Nos. 2284, 2288	Injunction halting export of all Mono water through March 30, 1990; water releases of 85 to 100 cfs for Rush Creek and 60 cfs down Lee Vining Creek to stabilize the lake's level.
1990	*California Trout v. Superior Court*, 218 Cal. App. 3d 187	SWRCB directed to amend LADWP's diversion licenses to restore streams to their 1940 status.
1991	State Water Resources Control Board	Order 91-04 amends diversion licenses in Owens Valley and Mono to comply with Fish and Game Code Sections 5946 and 5937.
1993	Environmental Protection Agency	Mono Basin in moderate nonattainment of federal PM-10 standards.
1994	State Water Resources Control Board	Decision D-1631 amends diversion rights to set permanent stream flows to public trust values in Mono Lake at a level of 6,392 feet.
1998	Great Basin Unified Air Pollution Control District	Adopted State Implementation Plan for dust control by LADWP on Owens Lakebed.
2004	*Sierra Club et al. v. City of Los Angeles*, Superior Court of Inyo County, SICVCV01-29768	Court enforces Memorandum of Understanding to release water in the Lower Owens River Project and to prepare a revised EIR.
2005	*Sierra Club et al. v. City of Los Angeles*, Superior Court of Inyo County	EIR challenged; court again orders release of water for the Lower Owens River with penalty of a halt in water exports.

SOURCE: Compiled by author.

representing about thirty-five years of conflict between the LADWP, local Inyo and Mono County governments, and environmental and recreation groups.[4]

In the end, after all of this effort in the courts and administrative agencies by both sides, Los Angeles lost water. In its 2005 Urban Water Management Plan, the Los Angeles Department of Water and Power stated that 40 percent of its historical aqueduct supply is now devoted to environmental enhancement in Mono and Inyo Counties. According to the report, this amounts to 166,000 acre-feet annually, or $83,166,000 a year with a present value of $1,427,052,576. As late as the period 1995 to 2000, the aqueduct provided 63 percent of the city's water. By

2001, this had dropped to 34 percent.[5] Chapters Six and Seven presented some of the specific costs borne by the city in dust mitigation in the Lower Owens River Project and in restoring stream flows and Mono Lake's level.[6]

THE COSTS OF JUDICIAL REALLOCATION OF WATER

When recreational, amenity, and environmental values of water were rising in the 1970s, there were growing demands to redirect aqueduct water to those purposes. In the face of these new values, some of the water obtained thirty to forty years earlier by Los Angeles for urban use most likely would have been devoted to them. The questions were: how much water would be redirected, how would the reallocation take place, and how quickly would it happen?

The path taken was litigation and uncompensated redistribution of the water. This approach may have been encouraged by the popular perception that Los Angeles's water rights were illegitimate because they were obtained by theft. According to this view, the price paid by the city to the farmers for their water in the 1920s was too low and in the 1970s did not reflect the environmental costs inflicted on the source region.

As we have seen, the LADWP did not steal the water rights from Owens Valley, and the value of environmental externalities and public recreational demands rose some thirty to forty years later as new demands appeared and as more water was diverted down the aqueduct. It seems, then, that an alternative, negotiated approach to secure some of Los Angeles's water might have been more effective for addressing legitimate new concerns.

As it was, and knowing that its ratepayers would bear most of the costs, the LADWP fought to retain its water rights. Local governments and environmental groups just as tenaciously pressed their demands. Positions hardened on both sides. Environmental advocates on the one side are described by some in the literature on Mono Lake as being driven, while on the other side the LADWP is depicted as being rigid and isolated.[7] These descriptions can also be read as representing increased polarization of each side's position and of the claims made by competing parties in an all-or-nothing setting.

As argued earlier, the issue of compensation and respect for property rights is more than one of equity. The long and complex pattern of conflict suggests that there were important efficiency effects of the enduring litigation of increased transaction costs, delayed response to environmental externalities, misallocated water, and more broadly, weakened property rights.

The alternative was to recognize Los Angeles's appropriative water rights and to use them as the basis for exchange. Negotiated exchange would seem to have been feasible between the LADWP and private groups or county governments for localized habitat restoration projects, reduced groundwater pumping on city lands, greater irrigation in Owens Valley and the Mono Basin, and perhaps, some enhanced flows in the Owens River and streams tributary to Mono Lake for trout fisheries. The amounts of money required to fund and organize those activities would have been sufficiently small. Other organizations have successfully negotiated these types of agreements. The Oregon Water Trust, for example, purchases water from farmers for instream flows.[8] The LADWP would have been the single seller, a circumstance that raises potential problems for market transactions, but the agency had a long history of other successful property sales in Owens Valley.

Where more substantial amounts of water were required to address public environmental concerns, such as refilling Mono Lake, the state of California (and the federal government, where appropriate) could have used bond funding or other revenue sources to acquire water rights from the LADWP.

Where no voluntary agreements on water transfers for public environmental or recreational uses were forthcoming, eminent domain with compensation would have been an option for government acquisition of water.[9] As discussed in Chapter Four, bilateral monopoly conditions (the LADWP as seller and government as buyer) could lead to a breakdown in bargaining. The limited threat of government intervention could change threat positions and promote negotiated settlement.

There are a number of advantages of acknowledging Los Angeles's appropriative water rights and using them as a basis for bargaining over water. One is that the parties would have gained a much more accurate assessment of the amount of water required to restore fish habitat, build riparian systems, raise Mono Lake's level, and control dust pollution. These are questions of science, engineering, and economics. Scientists, engineers, and on occasion economists were employed in the litigation. But the studies done by both sides of the conflict were part of an adversarial process to take or hold on to water. In their analyses, the parties were not disciplined by actually having to purchase water or to forgo the opportunity of sale. The expert reports were used to bolster each side's position for judicial rulings in the disputes. Hence it is not clear how useful they were in deciding on the best possible allocations. The information generated as part of public trust disputes may be particularly biased, as discussed below.

In negotiated exchanges in contrast, the buyer must offer the seller at least the latter's marginal reservation value. This requirement forces would-be purchasers to calculate how much water they want to buy and how much they are willing to pay for it. The process requires that both buyers and sellers evaluate the marginal values of the water they seek or hold. Under these less-confrontational circumstances, experts on both sides can more accurately determine the quantity of water that is truly needed. There is less incentive to distort or inflate the outcome. In Owens Valley, water would have traded whenever the marginal value of environmental uses exceeded those of urban consumers in Los Angeles. In this situation, the LADWP could then have used the revenues from water sales to secure alternative sources of supply.

Given the magnitude of some of the environmental effects in the source region, particularly the drawdown of Mono Lake, it seems clear that water was misallocated in light of new environmental concerns before the judicial and administrative allocation decisions outlined in Table 8.1 were made. But because there was no real balancing of relative water values in those decisions, there is no reason to conclude that the resulting water distribution between urban and environmental/recreational uses was socially optimal. In weighing alternative water allocations, environmental values are difficult to measure and quantify because they involve nontraded public goods. But economics has made substantial progress in valuing such assets, and contingent valuation was used to value various Mono Lake levels in the 1993 Environmental Impact Report.[10]

Even if the conditions for pure market exchange were not met for addressing new water demands, the use of eminent domain with just compensation would have provided more accurate information about relative values than did the approach taken. Since government would have had to purchase water using scarce revenues, there would have been more incentive to carefully determine the amount of water truly needed in the source region. At the same time, the LADWP would have had to justify the amount and value of the water it was diverting through the aqueduct.[11]

Another advantage of recognizing Los Angeles's appropriative water rights is that they could be the basis for subsequent water adjustments. Over time, environmental and recreational demands may grow; alternatively, urban demands may increase. With definite water rights, there is a legal framework for more-or-less routine contracting among the parties to redistribute more water as values change. The alternative seems to be more conflict of the sort that plagued the relationship between Owens Valley and Los Angeles.

A final and related advantage of acknowledging Los Angeles's water rights is that actions taken in major watersheds, such as Owens Valley and the Mono Basin, involving so much water, have profound effects throughout the West on the security of appropriative rights. Appropriative rights have been the basis for private investment in water distribution, storage, quality, and conservation since the nineteenth century. They have been the basis for recent private efforts to rehabilitate aquatic habitat in Montana and elsewhere in the Pacific Northwest, and more such investment is likely. Moreover, appropriative rights are the basis for the growing market trades outlined in Chapter One. There is a substantial economics literature on the importance of secure property rights for establishing incentives for investment, production, and trade.[12] Accordingly, protecting the appropriative rights doctrine is worth doing. At the same time, private rights may provide too few incentives for private investment in public goods provision, and in those cases, governments or nonprofit groups can acquire water through compensated exchange for those purposes without weakening the underlying rights structure.

THE COSTS OF THE PUBLIC TRUST DOCTRINE

In writing about the Mono Lake controversy, John Hart saw the 1983 public trust ruling of the California Supreme Court as offering a powerful way of balancing public and private demands for water by stressing its common property nature. Harrison Dunning and others have reached similar conclusions.[13] But for those who are concerned about water quality, supplies, and flexible allocation in the presence of growing scarcity, as are most westerners, caution is in order.

One problem is that the doctrine stresses water as a regulated common resource, and the experience with regulated commons often has not been that satisfactory.[14] Indeed, dissatisfaction with the past performance of centralized regulation of common resources has led to the adoption of more formal property arrangements. These include individual transferable quotas (ITQs) in common fisheries, tradable emission permits for air pollution reduction, and shares in unitized oil and gas fields.[15] In so doing, the resources involved have shifted from being considered "public" to more "private" in order to instill incentives for better stewardship and conservation. Where ITQs have been adopted, fishery stocks generally have rebounded and the value of the fisheries increased. Tradable emission permits have lowered the costs of achieving air quality standards. Oil-field unitization has brought important efficiency gains.[16]

Judicial expansion of the public trust doctrine in water appears to move in just the opposite direction. The doctrine as interpreted by California courts holds that bodies of water (navigable and nonnavigable) belong to the public (as common property) and that the government has a special and inescapable duty to protect them.[17] Accordingly, state agencies are charged under the trust to regulate allocation and use with adjustments made as public values change. This is a broader regulatory mandate: "Thus, the function of the Water Board has steadily evolved from the narrow role of deciding priorities between competing appropriators to the charge of comprehensive planning and allocation of waters. This change necessarily affects the board's responsibility with respect to the public trust."[18] It is not obvious why greater regulation of this "common" resource (water) would perform more effectively than has been the case in fisheries, air pollution, or oil pools.

Another problem with the doctrine is that it clearly weakens appropriative water rights. Their nonvested, usufruct nature is stressed, subject to continued reevaluation of their position vis-à-vis changing trust values. Use rights previously granted can be revoked without payment: "The foregoing cases amply demonstrate the continuing power of the state as administrator of the public trust, a power which extends to the revocation of previously granted rights," and "Once again we rejected the claim that establishment of the public trust constituted a taking of property for which compensation was required."[19] Vague, elastic notions of the public trust would create considerable uncertainty for any water rights holder. This problem would presumably apply not only to private water users, but to public ones as well, suggesting that current distributions for recreational or environmental protection, at a later date, could be found to be inconsistent with the public trust if values were to change. An advantage of secure property rights is that they can be a basis for routine reallocation through exchange. Public trust reallocations, however, would be made through regulatory rulings and not be compensable. The experience outlined in Table 8.1 suggests just how controversial and costly such reallocations could be.

A third problem with the public trust doctrine is that legal disputes brought under it may be more difficult to settle privately because of the broad legal standing it authorizes. When parties enter into litigation they weigh the expected benefits of trial versus settlement. Settlement typically is less costly than trial. If the net benefits of trial decline for one of the parties (higher cost, lower probability of winning, reduced damage expectations for plaintiff, higher damage expectations for defendant), settlement is more likely.[20] Under the public trust doctrine, however, settlement costs could be driven up as the number of potential plaintiffs grows. With extensive legal standing available for groups to challenge existing water uses as implied by expansive notions of the public trust, any settlement agreement with one party could be thwarted by the appearance of another plaintiff. Successive settlement negotiations would drive up settlement costs, and as they increased, disputes would be more likely to go to trial.

Public trust suits are also more likely to go to trial if the plaintiffs generate more information on their behalf at lower cost than the defendant. The contesting parties invest in lawyers, experts, and other resources to improve their chances of winning. Greater investments in litigation expenditures by one party relative to the other, all else equal, can increase the probability of success in the case.[21] With an ease of obtaining legal standing under the public trust doctrine, one could imagine entry by successive plaintiffs, each more extreme in its demands than the previous ones. If such plaintiffs are made up of ardent "true believers," it is possible that their labor costs for litigation expenditures would be lower than those for the defendant. Lower litigation costs, in turn, would lead to greater investment in litigation, a higher probability of winning, and a greater likelihood of costly trial. Moreover, judicial or regulatory decision-making based on information provided by the low-cost plaintiff may be biased because it would outweigh that provided by the higher-cost defendant.[22]

These are conjectures about the added costs of the public trust doctrine through lower settlement rates, more trials, and potentially lopsided judicial opinions regarding water allocation. No tests are provided here. There is some reason to believe, however, that the labor costs of litigation expenditures were lower for the plaintiffs in the Mono Lake case than for the LADWP. The literature on the episode describes the membership of local environmental groups as passionate, dedicated individuals, whose battle against the city to save Mono Lake was a crusade.[23] This discussion does not judge these efforts, but rather suggests that such dedication lowered the costs of litigation investments and increased the likelihood of trial.

CONCLUSION

This chapter has discussed the costs of judicial reallocation of water that has taken place in Owens Valley and the Mono Basin over the past thirty-five years. It has argued that the alternative of negotiated exchange based on Los Angeles's appropriative water rights to address environmental and recreational issues would likely have been more timely, less controversial, and less costly. The particular problems potentially introduced by the public trust doctrine were presented. Although the advocates of greater water for environmental uses in those regions and less for the Los Angeles Aqueduct may be satisfied with the outcomes of the actions described here, they may not be so pleased with the process. It was draining for all parties, and perhaps few would want it applied elsewhere where water conflicts are intensifying. Accordingly, the experience of Owens Valley and the Mono Basin can serve as motivation for different approaches to address the growing scarcity of water in the West.[24]

9 | CONCLUDING THOUGHTS ABOUT OWENS VALLEY AND WESTERN WATER

[Owens Valley] presents in such high relief a wide range of problems that exist generally in the management of water throughout California.

William Kahrl, Water and Power, *1982*

Owens Valley is important for western water both for what did *not* take place there and for what *has* taken place there. Los Angeles did not "steal" the water. But the contemporary view of Los Angeles's acquisition of land and water rights in the valley over seventy years ago is that the city stole them. According to this view, the Los Angeles Department of Water and Power moved across Owens Valley, secretly when it could do so, and bullying when it had to be, taking the water and the very livelihoods of the small farmers who lived there. They were left to face a colonial master who unfairly and uncaringly left the valley to dry up in order to meet the needs of extravagant water users in Los Angeles.

This is a very influential image, and it is used over and over again by those who oppose the transfer of water from rural areas for various reasons. It is a cautionary tale of what can happen to agricultural communities if their water is sold and exported. The problem is, as we have seen, that much of the story is inconsistent with the evidence.

If the price paid to farmers for their lands is evidence, there was no real theft. The LADWP had to buy the farms to get the water in Owens Valley. The sales negotiations between the agency and farm owners were complicated and often contentious. This is not because the farmers were holding on for dear life, but rather because they were so strategic in bargaining with the city. At a time of growing national agricultural depression, when farm land values were gradually declining, and in a region of marginal agricultural potential to begin with, the owners of the valley's small farms appear to have seen the opportunity to sell their lands to a rich city agency as a golden opportunity, one not to be missed. At the same time, however, they had a good sense of just how valuable their water was in Los Angeles, and they wanted as much of that bonanza as possible.

Those farmers who could, formed sophisticated sellers' pools to negotiate jointly with the LADWP. They enlisted the aid of the state legislature, the governor, and a sympathetic press to advance their bargaining positions for higher land prices. When negotiations faltered, they dynamited the aqueduct to capture the attention of the Board of Water and Power Commissioners, and they pressed hard with a public relations campaign. This was California's "little Civil War." And it worked. Those farmers who joined sellers' pools earned more than if they had remained in agriculture. Even those farmers who sold their properties individually did as well as or better than other Great Basin farmers.

Los Angeles did not want to buy the towns in Owens Valley in the 1920s and 1930s, because they brought no water. Political pressure forced it to do so, following claims of deteriorating property values due to the LADWP's farm purchases and the associated decline in the farm economy. California Board of Equalization data, however, reveal no such impact. Sellers' pools were formed in the towns as well, and the legislature demanded that the LADWP pay fair prices for the town properties. And the agency did. The LADWP bought the towns between 1929 and 1935, based on 1923 values, not later values that reflected the arrival of the Great Depression. A comparison of the mean prices paid for town properties with census values for rural nonfarm homes shows that Inyo County values were far above those in comparable Great Basin counties and indeed, in California as a whole.

The town-lot issue in Owens Valley is an example of the third-party effects that play such an important role in contemporary water transfers. They are difficult to assess, measure, and control, but they cannot be ignored. Many involve legitimate technological and pecuniary effects, whereas some may be pure rent-seeking. In any event, there typically is sufficient surplus from reallocating water to address valid claims. How these third-party effects are handled affects the sense of fairness of water transfers and the political viability of water marketing as a solution to the problem of water allocation in the American West.

Despite the farmers' organizational efforts and their success in selling their farms and their town properties, the implicit price they received for the water on their lands was too low, given its value to Los Angeles. Their little cartels were not strong enough to extract more of the returns from moving water from low-value farming in Owens Valley to much higher-value uses in Los Angeles. In comparison with what the LADWP had to pay to secure Colorado River water, Owens Valley water came much cheaper. And the farmers resented this outcome. They did not get as many of the gains from trade as they wanted. Nevertheless the legacy of the alleged "theft" of Owens Valley and Mono Basin water remains a powerful one in contemporary water debates. The controversy over Owens Valley shows that it is not likely to be feasible in the future to buy out an entire area to obtain its water. Transfers will involve a smaller share of a region's resources, leaving nonsellers concerned about any impact of the transaction on them. Monitoring for physical and pecuniary impacts will be an ongoing activity.

Let us turn to what has happened in Owens Valley and the Mono Basin since those sales. There have been conflicts between the LADWP and Owens Valley residents over the management of groundwater and the extent of pumping. When the agency purchased most of the land in the valley, it did so for access to both surface- and groundwater. The latter was to be the residual source of supply for the aqueduct. Environmental concerns rose after 1970 with completion of the second aqueduct and more intensive pumping to fill it.

Although the city internalized many of the physical externalities associated with groundwater pumping through its extensive property holdings, it did not internalize all of them. There were negative effects on wildlife habitat and riparian ecosystems. There also was criticism of related cutbacks in irrigation water that threatened the remaining agriculture in the valley. Many of these problems were localized enough that the agency, local residents, and the county government could have mitigated them through negotiation rather than resorting to costly litigation and regulation. The experience with pumping in Owens Valley underscores the importance both of effective groundwater management and of responding to local concerns about it. Timely and considered reaction by the LADWP would have been costly in terms of lost water, but these costs would likely have been less than those the agency had to bear later as its political position weakened.

Dust pollution off the dry Owens Lakebed was a larger externality, involving both the state and federal governments. Private contracting as a solution was therefore less feasible. In any event, it is unclear just how costly the pollution was or what the most effective response to it should have been. The debate continues. Meanwhile given the documented costs of the water and infrastructure required to mitigate the pollution, in the face of less well documented benefits of abatement, it is probable that the regulatory actions taken would not satisfy standard cost-benefit criteria. Unfortunately, the response to Owens Lakebed dust pollution was subsumed in, and distorted by, the broader battle between the LADWP and its antagonists over water.

The decline in Mono Lake's levels and the related impact on wildlife and amenity values after 1970 is the most serious physical externality caused by the

LADWP's water exports from the region. In the 1930s, when Los Angeles acquired Mono water rights, the effects of diversion on the lake were unanticipated. Further, the lake itself was not highly valued. Later, however, with increased water export and heightened environmental awareness, appreciation changed. The lake became seen as the center of a vital ecosystem worth protecting. Accordingly, Los Angeles would have had to reduce diversions.

The questions to be decided, however, were how much water to retain at the source and how to retain it. As we have seen, the course of action taken was controversial and prolonged, involving an extension of the public trust doctrine, an act that may have weakened water rights.

With rapidly growing cities, rising incomes, and increased environmental and recreational demands, there will be new and often conflicting claims on water and its distribution in the West. Many will involve out-of-basin exports. The observed response to these changes in Owens Valley and the Mono Basin does not seem to be a template for application elsewhere. In that sense, it is correct to conclude that "no more Owens Valleys" should take place.

NOTES

NOTES TO CHAPTER 1

1. 2005 Urban Water Management Plan, City of Los Angeles, Department of Water and Power (DWP), p. 3–3 (http://www.ladwp.com/ladwp/cms/ladwp 007157.pdf). For discussion of Owens Valley, see Libecap (2005a and 2005b).

2. Hanak (2003).

3. For contemporary discussion, see Howitt (1994); Northwest Economic Associates (2004).

4. There are the difficulties of distinguishing between legitimate claims of harm and pure rent-seeking actions, as well as of measuring the third-party effects. Equity issues are also discussed in National Research Council (1992) and Saliba (1987).

5. See for example, the discussion of battles over water in California's history at the Water Resources Center Archives (http://www.lib.berkeley.edu/WRCA/exhibit.html). Distributional issues are discussed by Carter, Vaux, and Scheuring (1994).

6. Hardin (1968); Libecap (1998). See also Glennon (1991, 2002).

7. For discussion of unitization of oil fields, see Libecap (1989, 93–114) and Libecap and Smith (1999). Kanazawa (2003) discusses early restrictions on groundwater transfers in California.

8. See Todd (1992) for discussion of Texas groundwater regulations.

9. Water markets compare unfavorably with those for land. For discussion of property rights and markets for land, see works by Ellickson (1993) and Rose (1998). A summary of water management and transfer issues, as well as

arguments for and against a greater market role is provided by Cohen (2005) and Glennon (2005).

10. Griffin and Boadu (1992, 274–75).

11. Dean E. Murphy, "Pact in West Will Send Farms' Water to Cities," New YorkTimes.com, October 17, 2003.

12. The data were drawn from the *Water Strategist* for water transfers in the twelve western states of Montana, Wyoming, Idaho, Washington, Oregon, California, Colorado, Utah, Nevada, Arizona, New Mexico, and Texas, from January 1987 through December 2005. These are the transfers where price data were included. Thirty-three had unknown origins or destinations, so 2,154 transfers are used in the analysis below. Other analyses of water transfer data are provided by Brown (2006), Howitt and Hansen (2005), and Loomis, Quattlebaum, Brown, and Alexander (2003).

13. Since the sample covers the years 1987–2005, prices were converted into real dollars to compare prices across years. All prices were converted into 1987 dollars using the Consumer Price Index—All Urban Consumers Average from the Bureau of Labor Statistics. The patterns shown in the table hold if median prices are used instead of mean values, although the differences are narrowed. Further, if transactions in the very active water market state of Colorado are removed, the price differences remain, but the number of observations is reduced.

14. This point is emphasized by Young (1986).

15. For discussion of the Colorado–Big Thompson Project and Northern Colorado Water Conservancy District rules and practices, see Carey and Sunding (2001).

16. Glennon, Ker, and Libecap (2007). There were 1,825 transfers for 5,533,394 acre-feet. The amounts reported here are yearly flows in the first year of the contract, as reported in the *Water Strategist*. Accordingly, the amounts understate the total water committed for transfer by sales and long-term leases.

17. Glennon, Ker, and Libecap (2007); 7,138,481 acre-feet were moved within agriculture, and 5,657,591 acre-feet within the urban sector, out of a total of 30,964,578 acre-feet transferred between 1987 and 2005.

18. Forty-six percent of agriculture-to-agriculture trades and 61 percent of urban-to-urban trades were sales.

19. Colby (1990, 1995).

20. For example, in Arizona under ARS 45-181 surface water is public; according to ARS 45-141 (A), water belongs to the public subject to appropriation and beneficial use. See Bokum (1996) for discussion of New Mexico; and Gould (1995, 94).

21. Wyoming Rules and Regulations, State Engineer, Chap. 1, Sec. 4. Water is property of the state. WY ST 41-3-101 says that individuals can have rights for beneficial use and that they can be sold with the land or detached from the land. Any change in place requires a no-injury rule (WY ST 41-3-104). See also the Wyoming Constitution, Art. 8, 1.

22. Gould (1995, 94); Simms (1995, 321).

23. Getches (1997, 11).

24. Sax (1990, 260); Gray (1994, 262); and Koehler (1995, 555).

25. Getches (1997, 74–189).

26. Getches (1997, 156–60).

27. Thompson (1993, 681); Glennon (2002, 14–21).

28. Getches (1997, 81).

29. See discussion of first possession by Epstein (1979); Rose (1985); and Lueck (1995, 1998). Prior appropriation rights are discussed by Getches (1997, 83).

30. Getches (1997, 8).

31. Getches (1997, 33).

32. Thompson (1993, 684).

33. Getches (1997, 251).

34. For discussion, see Glennon (2002, 8, 30, 210).

35. Provencher and Burt (1993); Glennon (2002). For similarities with oil pools, see Libecap (1989, 93–114).

36. Getches (1997, 247–53); Glennon (2002, 209–24).

37. Hanemann (2005, 9).

38. Getches (1997, 168–70).

39. Anderson and Johnson (1986); Johnson et al. (1981). Johnson et al. describe how specifying a property right in water in terms of consumptive use with options for third-party grievances can be an effective method for promoting transfers.

40. For discussion, see Gould (1995, 94) and Colby, McGinnis, and Rait (1989).

41. MacDonnell (1990, 11). In most western states, local water users can change the point of diversion if no harm is caused.

NOTES TO CHAPTER 2

1. The notion of "hydrocolonialism" is used in Carl Nolte, "A Woeful Tale of Stolen Water, Quaking Earth," *San Francisco Chronicle*, September 4, 1986.

2. Haddad (2000, xv).

3. Kahrl (1982, 2000a, 2000b). The 2000a and 2000b articles are reprints of articles that appeared in the 1976 spring and summer issues of *California*

Historical Quarterly. See also, Libecap (2005) for discussion of the Owens Valley water transfer.

4. Kahrl (1982, 38–39).

5. Kahrl (1982, 40–79, 92–103, 184; 2000a, 241, 247).

6. Kahrl (1982, 85–91, 442; 2000a, 243).

7. Kahrl (2000a, 242).

8. Kahrl (2000b, 263–64).

9. Kahrl (2000b, 259–61).

10. Kahrl (1982, 90–94, 132, 183–94, 227; 2000a, 245).

11. Kahrl (2000a, 239). Variants of this theme are found in other books on Owens Valley, including Wood (1973, 5–8), and Ewan (2000, 52, 130–32, 139, 142, 152). More balanced views of Owens Valley, however, are in Sauder (1994, 109, 124–34, 146, 151–64); Hoffman (1981, xviii–xiv); and Nadeau (1950, 126–28).

12. Getches (1993, 535).

13. Davies (1999, 113–14).

14. Lyon (2002, 134–35).

15. Rothman (2002, 216–17).

16. Sawyer (1997).

17. "California's Liquid Asset. Cities Want to Buy Imperial Valley's Water," *Washington Post*, November 3, 1985.

18. John M. Broder, *New York Times*, August 8, 2004, 14.

19. Wheeler (2002, 105).

20. Ostrom (1971, 449); Haddad (2000, xv); Gray, Driver, and Wahl (1991, 979).

21. Hanak (2003, 5, 123); Hanak and Dyckman (2003, 495); Beck (2000, 139); Tarlock (2000, 102).

22. Ralph (2003).

23. Thompson (1993, 733 n. 262, notes that residents of the Imperial Irrigation District opposed agreement with Metropolitan Water District in order to avoid turning it into another Owens Valley.

24. See Wood (1973, 8); Reisner (1986, 60–107); and Ewan (2000, 42).

25. See Nadeau (1950, 126–28); Hoffman (1981, xviii–xiv); Vorster (1992); Walton (1992, 192–97); and Sauder (1994, 124–34, 151–64).

26. Ralph (2003, 906 n. 23).

27. Ralph (2003, 919).

28. Hanak and Dyckman (2003, 495).

29. *Baldwin v. County of Tehama*, 36 Cal. Rptr. 2d 886 (Cal. Court. App. 1994).

30. Hanak and Dyckman (2003, 511, 518).

31. Lyon (2002, 135–36).

32. Rothman (2002, 216–17).

33. Sawyer (1997, 322–24).

34. Sax (1994, 15).

35. Nichols and Kenny (2003, 425).

36. Nichols and Kenny (2003, 425–26).

37. Quoted in Elliot Diringer, "Owens Valley: Victim of Water Wars," *Oklahoma City Journal Record*, June 9, 1993.

38. Kahrl (2000b); Reisner (1986, 52–103).

39. A search of the Westlaw ALLNEWS database for the terms "Owens Valley" and "steal" or "stole" or "rape" yielded a total of eighty-seven relevant articles, almost half of which (forty-two) were published by either the *Los Angeles Times* or the *Los Angeles Daily News*. While the legend of Owens Valley has been taken up by the press in other, primarily western, states and even by some international papers, the California press and Los Angeles press, in particular, remain active disseminators of the story. An expanded search, adding the term "Mulholland" to the string, yielded thirty-six relevant articles published between 1985 and 1990, thirty-two of which were printed in the *Los Angeles Times*, one in the *Los Angeles Daily News*, and two in other California papers. The sole article in the search not printed in a California source was an Associated Press story.

40. Robert Reinhold, "Accords Reached over Water, L.A. Yields Some Mono Lake Rights," *Los Angeles Daily News*, September 25, 1989.

41. "For Shame, San Francisco," editorial, *Los Angeles Times*, September 2, 1986.

42. "California's Future Is Here," editorial, *Los Angeles Times*, July 22, 2001.

43. "It's Still a Desert," editorial, *Los Angeles Times*, October 20, 2003.

44. See the interview with S. David Freeman, the Department of Water and Power's general manager, referencing theft of Owens Valley water in Richard Nemec, "Viewpoint," *Los Angeles Daily News*, March 28, 1999, and reference to theft of Owens Valley water in Paul Byrnes, "Fear and Loathing in Los Angeles—Just Another of Its Selling Points," *Sydney Morning Herald*, April 15, 1999.

45. *Economist*, July 19, 2003, p. 15.

46. "A Truce, Perhaps a Treaty," editorial, *Los Angeles Times*, February 6, 1985.

47. Carl Nolte, "A Woeful Tale of Stolen Water, Quaking Earth," *San Francisco Chronicle*, September 4, 1986.

48. John Kamman, "Hanging Rural Arizona Out to Dry," *Arizona Trend*, July 1, 1987.

49. Elliot Diringer, "Owens Valley: Victim of Water Wars," *Oklahoma City Journal Record*, June 9, 1993.

50. "Water: Up for Grabs?" editorial, *Los Angeles Times*, April 10, 1985.

51. Cass Peterson, "California's Liquid Asset. Cities Want to Buy Imperial Valley's Water," *Washington Post*, November 3, 1985.

52. "Sounds Like Old Times," editorial, *Los Angeles Times*, November 28, 1987.

53. "Water 'Bank' Proposed to Assuage Supply Shortage," *Los Angeles Times*, November 26, 1989.

54. Clyde Weiss, "Salton Sea Plan Could Hurt State Water Supplies," *Las Vegas Review-Journal*, March 13, 1998.

55. "Environmental Consequences of L.A.'s Search for Water," *Dallas Morning News*, October 1, 1989.

56. "L.A. Loses Its Grip," editorial, *Press Enterprise* (Riverside, Calif.), November 29, 2001.

57. Mark Trahant, "Water Cry: 'Share It,' not 'Take It'," *Seattle Times*, April 18, 1999.

58. Mike Davis, "House of Cards," Sierra 80 (6), November 21, 1995.

59. Kahrl (1982, 38–39).

60. "Speculating in Water," *Los Angeles Times*, September 10, 1989.

61. Rocky Baker, "Collaboration Key to Balancing Market Needs," *Idaho Statesman*, July 18, 2003.

62. Paul Rogers, "Failure of Transfer Deal Sets Stage for Water Fight," *San Jose Mercury News*, December 11, 2002.

63. Penelope Purdy, "Water Wars in the West," *Denver Post*, September 5, 2000.

64. John Woestendiek, "War over Water Being Waged in Colorado Valley," *Tulsa World*, January 6, 1991.

65. The arguments that follow are drawn from Libecap (2005).

66. See Glaeser, Kallal, Schenkman, and Shleifer (1992); Rauch (1993); and Feldman and Audretsch (1999) for discussions of the positive economics of cities.

67. See Jacobs (1984); Kim and Margo (2003, 40–41); and Glaeser, Kahn, and Rappaport (2000).

68. See summary of historical U.S. patterns in Kim and Margo (2003).

69. For technology externalities in economic growth, see Mills (1967); Henderson (1974); and Kim and Margo (2003, 38–40).

70. Ethier (1982).

71. See Helseley and Strange (1990) and Glaeser, Kolko, and Saiz (2001).

72. Kim and Margo (2003, 27–34).

73. Seth Hettena, "Parched City Draining Rural Livelihood; Colorado Farmers Bemoan End of Era but Sell Water Rights to Remain Solvent," *Washington Post*, February 1, 2004 (emphasis added).

NOTES TO CHAPTER 3

1. An important discussion of William Mulholland, chief engineer and architect of the Owens Valley project, is provided by Catherine Mulholland (2000). For additional analysis of the Owens Valley water transfer, see Libecap (2005).

2. The Los Angeles Board of Water Commissioners in its *Annual Report* for 1904 noted that local sources beyond the Los Angeles River were too limited to be of much help and that the city would have to find more remote supplies of water (1904, 25).

3. Mean precipitation for Los Angeles, 1921–2002, from www.nwsla.noaa .gov/climate/data/cqt_monthprecip_cy.txt; mean for Chicago, 1871–2003, from home.att.net/~chicago_climo/CHIPRCP.gif.

4. Ostrom (1953, 23) provides data on the various sources of water for Los Angeles, 1920–50. See Los Angeles Department of Public Service (1916, 32) for a discussion of the menace of water shortage in Los Angeles.

5. Miller (1977, 49–50). Phillips (1967, 16–17) provides smaller, but still very large figures of annual runoff of 510,000 acre-feet for the Owens basin and aquifer storage of 12 million acre-feet.

6. Hyde Forbes, Engineer and Geologist, "To the Owners of Lands Involved in the Aberdeen and Eight Mile Ranch Cases, Owens Valley, California and Their Counsel," File M0 BK G4551 E 5-3, Water Resources Research Center Archives, U.C. Berkeley. See also discussion of the geology and hydrology of the valley in J. C. Clausen, Engineer, "Report of the Owens Valley, California," November 1904, Eastern California Museum, Independence, Calif.

7. 1920 U.S. Census.

8. The 1925 U.S. Census of Agriculture says that there were 144,182 total acres of farmland in Inyo County in 1925; 140,029 in 1920; and 110,142 in 1910 (U.S. Bureau of the Census, Department of Commerce, U.S. Census of Agriculture, 1925, Western States, "California," Table 1, 1926). Irrigation estimate from Miller (1977, 53). Estimates of the amount of land in irrigation vary widely, up to 63,636 acres, but this figure is an outlier and seems unlikely.

9. See Miller (1977, 84–141) for discussion of water rights issues as they pertained to Owens Valley.

10. Charles H. Lee and S. B. Robinson, "Use of Water for Irrigation in Owens Valley in Connection with the Supply of the Los Angeles Aqueduct," February 1910, pp. 2–5, MS 1 98.17, Lee Folder, Water Resources Research Center Archives, U.C. Berkeley.

11. Conkling (1921, 11–12).

12. "Recent Purchases of Water in Owens Valley by City of Los Angeles," November 1923, Cope Rand Means Co., Engineers, San Francisco, Lee Folder,

MS 7611 98246, p. 7, Water Resources Research Center Archives, U.C. Berkeley (hereafter cited as "Recent Purchases of Water in Owens Valley," Water Resources Research Center Archives, U.C. Berkeley).

13. In the Imperial Irrigation District, use ranges from 5.6 to 6.6 a.f./acre; in the San Joaquin Valley it ranges from 2.5 to 6 a.f./acre. Data provided by Ellen Hanak of the Public Policy Institute of California. Kahrl (1982, 257) comments on the "poor" water management practices of farmers in the valley, a condition that suggests the existence of excess water.

14. Miller (1977, 53–55); U.S. Census; and Barnard and Jones (1987). The data are average farm size in Inyo County and the means of the average farm sizes for Churchill, Lyon, Douglas, and Lassen Counties. Similarly, the data are the average value of farm production per farm for Inyo and the mean of the averages for the other four counties. The agricultural potential of Owens Valley generally is exaggerated in the literature. For instance, see Kahrl (1982, 38).

15. Charles H. Lee, "Statement of Lands in Inyo and Mono Counties Owned by Department of Public Service of the City of Los Angeles Prior to January 1923," January 1924, Lee Folder, MS 7611 98L4a, Water Resources Research Center Archives, U.C. Berkeley (hereafter cited as Lee, "Statement of Lands," Water Resources Research Center Archives, U.C. Berkeley).

16. Sauder (1994, 113).

17. Ostrom (1953, 14); Nadeau (1950, 21–23).

18. Los Angeles Board of Water Commissioners, *Annual Report of the Board of Water Commissioners for the Year Ending November 30, 1905*, including "Report on Water Supply" by William Mulholland, Superintendent, and Lippincott & Parker, Consulting Engineers, Los Angeles (1906, 4).

19. Hoffman (1981, 99).

20. Resolution, Correspondence file, March–November 1905, May 22, 1905, Tape GX0001, LADWP Archives. City agrees to pay $450,000 for sixteen miles of Owens River frontage, 22,000 acres of land and an easement to 2,684 acres as a reservoir site. L.A. Board of Water Commissioners, *Annual Report* (1906, 5). See also Ostrom (1953, 13–14).

21. Lee, "Statement of Lands," Water Resources Research Center Archives, U.C. Berkeley,

22. Ostrom (1953, 118–19).

23. See discussion of the purchase of private water works, for example, in Los Angeles Board of Public Service Commissioners (1915, 7). See also Van Valen (1977).

24. Hoffman (1981, 141–54); Kahrl (1982, 90–103); and Ostrom (1953, 149–54) describe the bond issues in 1907, sources of opposition, outrage over land

speculation in the San Fernando Valley, conflict over annexation, and disputes between the city and private power and water companies over compensation for their properties. Bond issue data are provided in L.A. Board of Water Commissioners, *29th Annual Report* (1930, Exhibit V).

25. Short summaries of the scandal over insider land purchases are provided in Ostrom (1953, 58, 149–51); Kahrl (1982, 195); and Nadeau (1950, 29–41).

26. Hoffman (1981, 157–65); Ostrom (1953, 152).

27. Nadeau (1950, 29) noted that property values in much of Los Angeles doubled in price in 1905, when the Owens Valley project was announced.

28. Kahrl (1982, 170); Hoffman (1981, 154); Nadeau (1950, 32–35); and Ostrom (1953, 149–51) describe early water distribution and speculation in the San Fernando Valley.

29. Sauder (1994, 122); Ostrom (1953, 148). Quinton, Code, and Hamlin outlined the proposed distribution of Owens Valley water (Los Angeles Board of Public Service Commissioners, *Tenth Annual Report*, 1911).

30. See *Tenth Annual Report of the Board of Public Service Commissioners for the Year Ending June 30, 1911*; includes "Report upon the Distribution of the Surplus Waters of the Los Angeles Aqueduct," by J. H. Quinton, W. H. Code, and Homer Hamlin, which calls for water to be supplied to areas that agree to be annexed to the city; eliminates legal questions and requires that those districts pay for water distribution (p. 58).

31. Besides the movie *Chinatown*, there is extensive discussion of the land boom in the San Fernando Valley with Owens Valley water. See Nadeau (1950, 29–32, 41, 62–68); Hoffman (1981, 154–73); Kahrl (1982, 195); Ostrom (1953, 149–63); Ostrom (1971, 448–49).

32. I provide these figures only to illustrate the magnitude of the large gains in Los Angeles County. The $113.7 million is calculated from the 757,985 acres of agricultural land in 1910 as listed in the census multiplied by a possible and plausible increase in land value of $150 per acre as water became more certain and as Los Angeles's urban population moved into farming areas.

33. Hoffman (1981, 50–71).

34. Quoted in Hoffman (1981, 131).

35. This apparent deception is stressed in all historical discussions of the Owens Valley controversy; for instance, see Ostrom (1953, 116–18).

36. Details on Lippincott's alleged conflict of interest are provided by Hoffman (1981, 68–79, 136–41) and Kahrl (1982, 39–79, 85–140).

37. Conflict of interest is charged by Nadeau (1950, 28–31). Miller (1977, 66–79) reported concerns about Owens Valley's high water table, need for drainage, and high elevation. Pisani (1984, 302) describes the problems with

Owens Valley relative to other more promising sites in the West. See also Hoffman (1981, 47–60).

38. This is discussed further in Chapter Four. See also, for example, evidence that there was apparent lingering resentment on this issue in a letter from W. W. Yandell and Ione Seymoure of the Farmers Ditch Company to the Grand Jury of Inyo County regarding Los Angeles's purchase of McNally Ditch, September 22, 1924, Town Properties file, Tape GX0007, LADWP Archives. See also discussion of the legacy of the Reclamation Service project's cancellation by Hoffman (1981, 105, 112–19).

39. The engineer Hyde Forbes commented on the need for drainage to improve the effectiveness of irrigation in a court case later regarding another issue, "General Notes: Testimony of Hyde Forbes re: Aberdeen and Eight Mile Ranch Cases, Owens Valley, 1925," p. 54, Lee Folder, MS 7611 98L4a, Water Resources Research Center Archives, U.C. Berkeley.

40. Kahrl (1982, 39, 43–79, 105–47); Hoffman (1981, 47–90).

41. The aqueduct was viewed as second only to the Panama Canal in its engineering achievement. Nadeau (1950, 45–60); Ostrom (1971, 447–48); Los Angeles Department of Public Service (1916, 17–29).

42. Osborne (1913).

43. Nadeau (1950, 43–59). Miller (1977, 150) provides different numbers, but in any event, this was a large project.

44. Hoffman (1981, 145–53).

45. http://wsoweb.ladwp.com/aqueduct/default.htm. The second aqueduct, opened in 1970, added 290 cubic feet per second (cfs) and was 137 miles long.

46. Bateman et al. (1978, 195).

47. Ostrom (1953, 2); Conkling (1921, 8). The cubic feet per second are translated to annual acre-feet by multiplying by 1.98 and 365 days.

48. Ostrom (1953, 14, 23); Miller (1977, 158).

49. Ostrom (1953, 23); see also Kahrl (1982, 227–30).

50. Lee, "Statement of Lands," Water Resources Research Center Archives, U.C. Berkeley. Deflator, University of Michigan Library Historical Consumer Price Index, 1880–1998 (http://www.lib.umich.edu/govdocs/historiccpi.html).

51. Conkling (1921, 8).

52. Nadeau (1950, 64–67).

53. Ostrom (1953, 120–48); Hoffman (1981, 172–74), and Sauder (1994, 118–19, 137–39). Kahrl (1982, 239–50) discusses the issue, but does not believe that the reservoir would have saved irrigation in Owens Valley (p. 252).

54. Sauder (1994, 137).

55. See Kahrl (1982, 222–24).

56. Conkling (1921, 5). See statement in Owens Valley Irrigation District File, Tape GX0003, LADWP Archives. There are many estimates of irrigated acreage, but 35,000 is consistent with the 1920 Agricultural Census.

57. Ostrom (1953, 15–16) describes the drought of the early 1920s.

58. Kahrl (1982, 233, 269–70).

59. See discussion in "Owens Valley Situation, Synopsis," Correspondence file, January–March 1929, Tape EJ00086, LADWP Archives.

60. Miller (1977, 159–63) says that 1923–24 was the driest in recorded history in California.

61. The LADWP Archives have extensive files on pumping cases. See, for example, a letter regarding ongoing litigation over the Longyear ranch, W. B. Mathews file, Tape EJ00086; memo from board counsel Mathews regarding damage claims, December 16, 1924, Litigation file, Tape GX0002; and 1930 documents in which the board expresses worry about jury awards, Yandell Case file, Tape GX0005.

62. Letter in 1925 outlining dispute over Longyear properties, W. B. Mathews file, Tape EJ00086, LADWP Archives.

63. Letter from board to two landowners, C. P. Crowell and S. F. Zombro, reporting on the status of land purchases in Owens Valley, "Tabulation Showing Status of Ranch Land Purchases Made by the City of Los Angeles in the Owens River Drainage Area from 1916 to April 1934," prepared in Right of Way and Land Division by Clarence S. Hill, Right of Way and Land Agency, compiled by E. H. Porter, April 16, 1934, Sale of Lands file, Tape GX0004; LADWP Archives.

64. Miller (1977, 44–56).

65. For a discussion of ditch companies, see Israelsen, Maughan, and South (1946).

66. "Classified Acreage of Lands under Ditch, Bishop–Big Pine Region of Owens Valley, Based on Surveys by City of Los Angeles, 1922 to 1926," Miscellaneous file, Tape GX0004, LADWP Archives. See also By-laws of the McNally Ditch Company, McNally Ditch file, Tape GX0008, LADWP Archives.

67. Estimates of the amount of farmland vary. Besides the census, see "Classified Acreage of Lands under Ditch . . . 1922–1926," Miscellaneous file, Tape GX0004, LADWP Archives, which suggests that the ditches covered 77 percent of all farmland, or around 63,000 acres. Estimates of irrigated acreage are from Miller (1977, 53). Irrigated acreage of 70,500 is reported in "Recent Purchases of Water in Owens Valley," Water Resources Research Center Archives, U.C. Berkeley. This is much larger than other estimates and seems too large.

68. Water claims were usually for 180 days of use, reflecting the seasonality of agriculture in the region, so the 47 percent use of the claimed water is not totally an artifact of seasonality. See discussion of water claims in "Recent Purchases of Water in Owens Valley," p. 6, Water Resources Research Center Archives, U.C. Berkeley.

69. Statement to Mayor's Advisory Committee Prepared by the Special Owens Valley Committee of the Board of Public Service Commissioners," December 16, 1924, Special Owens Valley Committee file, Tape GX0004, LADWP Archives.

70. "Recent Purchases of Water in Owens Valley," p. 6, Water Resources Research Center Archives, U.C. Berkeley.

71. "Summary of Report on the Water Supply for the City of Los Angeles and the Metropolitan Area," Louis C. Hill, J. B. Lippincott, A. L. Sonderegger, Board of Engineers, August 14, 1924, pp. 1–13, Condemnation Proceedings file, Tape GX0001, LADWP Archives. Among the various options, the report outlined acquiring all of the water rights in Owens Valley, construction of a large storage reservoir in Long Valley, and extension into the Mono Basin.

72. Resolution of the Board of Public Service Commissioners of the City of Los Angeles, Correspondence file, October 1924, Tape EJ00087, LADWP Archives. This resolution agreed to leave 30,000 acres in irrigation.

73. "Recent Purchases of Water in Owens Valley," pp. 1–2, Water Resources Research Center Archives, U.C. Berkeley.

74. "Recent Purchases of Water in Owens Valley," Water Resources Research Center Archives, U.C. Berkeley. McNally Ditch (thirty farms or parts of farms for $918,500) and Big Pine Canal (twenty-five farms, 4,600 acres, $1.1 million, or $239/acre).

75. Nadeau (1950, 74–75).

76. Nadeau (1950, 71–74); Kahrl (1982, 283).

77. "Recent Purchases of Water in Owens Valley," Water Resources Research Center Archives, U.C. Berkeley.

78. W. F. McClure, State Engineer, "Owens Valley—Los Angeles, Report Made at Request of Governor Friend Wm. Richardson, Following the Opening of the Alabama Hills Waste Gates of the Aqueduct by the Valley People on November 16, 1924," December 26, 1924, p. 8, Eastern California Museum.

79. W. R. McCarthy, District Engineer, J. C. Clausen, Consulting Engineer, "Report on the Owens Valley Irrigation District, Inyo County, California," Report 36, March 20, 1923, Bishop, p. 1, Eastern California Museum.

80. Details on the formation of the Owens Valley Irrigation District are found in W. R. McCarthy file; Owens Valley Irrigation District Financial

Records file, "Petition for the Formation of an Irrigation District"; and Financial Records file, Tape GX0003, LADWP Archives.

81. Their claims amounted to 180,000 a.f. of a total of 234,000 a.f. of agricultural diversion. See Hoffman (1981, 176–79); Kahrl (1982, 277).

82. Kahrl (1982, 280).

83. McCarthy and Clausen (1923, 26).

84. See letter from Yandell and Seymoure to the Grand Jury of Inyo County, September 22, 1924, LADWP Archives.

85. Kahrl (1982, 278–79). Kahrl claims that holdouts were left without water, but this seems unfounded. Those ditch members who did not sell still received their share of ditch water. Whether they bore other costs, such as higher ditch maintenance costs, as is sometimes alleged, would be another issue. This issue is addressed below.

86. Kahrl (1982, 281).

87. The role of the purchase of the McNally and Big Pine Ditches in thwarting the effective organization of the Owens Valley Irrigation District, which would have united all of the sellers' pools, is described in the letter from Yandell and Seymoure to the Grand Jury of Inyo County, September 22, 1924, LADWP Archives; and "Percentage of Water Stock Owned by City of Los Angeles in Private Ownership in the Following Ditch Companies," Ditches file, Tape GX0001, LADWP Archives. See also Kahrl (1982, 279); Nadeau (1950, 95); Sauder (1994, 140–43).

88. The difference between the sellers' pools and OVID was that the latter was a recognized organization of the state with the ability to sell bonds and levy assessments. Water rights were transferred from district members to the district itself. The pools were more informal, organized and led by leading property owners. They also were much smaller. The pools did not hold water rights independently of landowners.

89. *Tenth Annual Report of the Board of Public Service Commissioners for the Year Ending June 30, 1911* (1911, 43).

90. Miller (1977, 164); Mrs. G. L. Wallace and Mrs. J. H. Stofflet in "Transcript of Proceedings, August 13, 1926, Ladies Committee to Board of Water and Power Commissioners," Owens River and Big Pine Canal file, Tape GX0003, LADWP Archives. Delameter (1977, 35) claims that Los Angeles kept water flowing in ditches even when it was the majority shareholder.

91. Hoffman (1981, 179–81).

92. Delameter (1977, 35) states that there was strategic use of checkerboarding by Los Angeles.

93. As argued in the next chapter, it was in the board's interest to follow such a strict pricing rule. The board was anxious to complete purchases as

quickly as possible for water-bearing lands. See Telegram from W. A. Lamar, Board Land Agent, to W. B. Mathews, August 14, 1925, Tape GX0004, LADWP Archives. Absent detailed maps of property locations, it is hard to determine whether there was deliberate checkerboarding.

94. Letter from attorney John Neylan to W. B. Mathews of the Legal Department October 8, 1926, complaining that although his client had not attempted to gouge the city and had stayed away from pool, she had not heard about her offer to sell for three years, Lone Pine file, Tape GX0003, LADWP Archives.

95. Wood (1973, 30–37); Ostrom (1953, 121–27).

96. Letter from William Kelso to George Watterson, December 24, 1924, Miscellaneous file, Tape GX0002, LADWP Archives.

97. Walton (1992, 171).

98. McClure, "Owens Valley–Los Angeles, Report" (1924, 10, 2).

99. *Literary Digest*, December 6, 1924, pp. 13–14. Letter from land agent John Martin to William Mulholland, May 9, 1924, claiming that the dynamiting was an effort to force the city to buy at "exorbitant prices," Tape GX00086, LADWP Archives.

100. "The Dynamite Holdup," Statement by the Board of Water and Power Commissioners, Miscellaneous file, Tape GX0001, LADWP Archives.

101. Miscellaneous file, Tape GX0004, LADWP Archives.

102. See also the summary of critical press articles in McClure, "Owens Valley–Los Angeles, Report" (1924, 46–101).

103. Letter from John A. Merrill to Board of Public Service Commissioners, August 15, 1927, Correspondence file, June to September 1927, Tape EJ00086, LADWP Archives.

104. *Hollywood Daily Citizen*, editorial, Clippings file, n.d. [probably late 1929], Tape GX0001, LADWP Archives.

105. Charges of conspiracy among Owens Valley landowners to overcharge the city were made by the Los Angeles Municipal League in its *Bulletin* 5 (3), October 31, 1927, W. W. Watterson Correspondence file, Tape GX0005, LADWP Archives. On the other hand, the "complete lack of confidence on the part of many of the Valley people toward the present administrative officers of the Departments" is emphasized by the Special Owens Valley Committee of the Board of Public Service Commissioners, December 16, 1924, Mayor's Advisory Committee file, Tape GX0003, LADWP Archives.

106. Ostrom (1953, 59–62). In 1925, under a new city charter, the board was required to cover all interest and principal charges on outstanding bond debt from its revenues, not from taxpayers, "Water Bonds" advertisement, Power and Water Bond Election file, August 31, 1926, Tape DS0010, LADWP Archives.

107. Kahrl (1982, 17) discusses the makeup of the board.

108. See L. A. Board of Water and Power Commissioners, *Annual Report* (1933, "Roll of Commissioners").

109. Hoffman (1981, 44–45).

110. See Hoffman (1981, 35–44) and Ostrom (1953, 59–64).

111. Kahrl (1982, 287).

112. Ostrom (1953, 124–25).

113. "Why Not Settle the Owens Valley Trouble?" *Municipal League Bulletin* 4 (11), July 30, 1927. Other press attacked the city's actions; see a series in the *Sacramento Union*, March 28–April 2, 1927, all in Miscellaneous file, Tape GX0004, LADWP Archives; telegram from the Owens Valley Protective Association to the Mayor, May 13, 1927, Reparations file, Tape GX0004, LADWP Archives.

114. Kahrl (1982, 304–6).

115. Nadeau (1950, 96).

116. Miller (1977, 164 n. 20).

117. Hoffman and Libecap (1991).

118. See letter from John Bartlett, First Assistant Postmaster General, to D. Parks, Chief Accountant, Department of Water and Power, regarding gross receipts at post offices at Big Pine and Bishop, 1921–26, and other conditions, February 1, 1927, Condemnation Proceedings file, Tape GX0001, LADWP Archives. Kahrl (1982, 169).

119. Statement by Paul G. Hoffman, Studebaker seller, April 15, 1927, that automobile registration was up in Owens Valley. Condemnation Proceedings File, Tape GX0001, LADWP Archives. Kahrl (1982, 144, 297) claimed that the economy had been devastated but presented the city's counterclaims.

120. For entries by the committee and board resolutions about economic conditions in the valley, see Special Owens Valley Committee file, Tape GX0004, LADWP Archives.

121. Walter Packard, Owens Valley Committee File, "Report on Proposed Settlement of a 3,000 Acre Tract of Land in the Big Pine District of Inyo, County California," December 3, 1924, Special Owens Valley Committee file, Tape GX0004, LADWP Archives.

122. Special Owens Valley Committee File, Tape GX0004, LADWP Archives.

123. Letter from W. F. McClure as State Engineer to Board of Public Service Commissioners, December 14, 1924, W. F. McClure file, Tape GX0003, LADWP Archives. Ostrom (1953, 123–27).

124. Hoffman (1981, 176–202).

125. Statement to Mayor's Advisory Committee Prepared by the Special Owens Valley Committee of the Board of Public Service Commissioners,

December 16, 1924, Special Owens Valley Committee file, Tape GX0004, LADWP Archives.

126. Hoffman (1981, 284–301). Report of the Special Owens Valley Committee to the Board of Directors of the Los Angeles Chamber of Commerce, August 30, 1927, Correspondence file, June–October 1927, Tape GX0086, LADWP Archives.

127. Chapter 109 of the Statutes and Amendments to the Codes of California, 1925; 1925 Laws of California, 251, enacted and approved by the governor, May 1, 1925.

128. Most opposition was shown through abstentions. In the Assembly, there were only three no votes, all from Los Angeles, but fifteen abstentions. Southern California, where opposition was greatest, included Los Angeles, Santa Barbara, Riverside, San Bernardino, San Diego, Orange, Imperial, and Ventura Counties. Overall representation in the legislature still heavily favored northern California, with fifty-six of the eighty members of the Assembly and twenty-eight of the forty members of the Senate. Legislation: 46th Legislature, Laws of 1925, p. 251; 46th Legislature, Senate Bill 757 (SB757), Chapter 109, enacted and approved May 11, 1925, with list of California Senate and House votes on the law. *The Senate Journal*, April 13, 1925, p. 1441.

129. "Facts Concerning the Owens Valley Reparation Claims for the Information of the People of California," Department of Water and Power, City of Los Angeles, Miscellaneous file, Tape GX0004, LADWP Archives. The totals given are 548 claims for $2,813,355, close to those listed in the table.

130. Hoffman (1981, 253); $6.6 million was used for Owens Valley and the rest for land purchases in Mono County.

131. Ostrom (1953, 126).

132. L.A. Board of Water and Power Commissioners, *Thirty Fourth Annual Report* (1935, 126). The LADWP acquired 21,915 acres from forty-one landowners in the Mono Basin for $4,826,949.

NOTES TO CHAPTER 4

1. For analysis of the capital gains of land sales and wealth accumulation on the frontier, see Ferrie (1994); Galenson and Pope (1989).

2. "Tabulation Showing Status of Ranch Land Purchases Made by the City of Los Angeles in the Owens River Drainage Area from 1916 to April 1934," prepared in Right of Way and Land Division by Clarence S. Hill, Right of Way and Land Agency, compiled by E. H. Porter, April 16, 1934, Tape GX0004, LADWP Archives (hereafter referred to as "Porter file").

3. Coase (1937, 1960); Barzel (1982); Dahlman (1979); Demsetz (1964, 1968); and Williamson (1979, 1981). Useful summaries of transaction-cost issues and concepts are in Allen (2000) and Eggertsson (1990).

4. Haddock and McChesney (1991) examine bargaining in a different setting, but they summarize some of the issues that raise bargaining costs when discovery of the terms of trade is not quick or easy and there is strategic noncooperation. See also Kennan and Wilson (1993) for a general summary of bargaining issues.

5. Heterogeneous farmland, of course, violates a condition for perfect competition. The setting presented here, however, is one of intense competition among multiple sellers and buyers.

6. Williamson (1975, 238–47); Blair, Kaserman, and Romano (1989).

7. One might have expected middlemen to emerge to arbitrage farm prices, but they are not apparent in the record. One reason might be that farmers were not interested in dealing with them, but preferred to deal with the board itself, where they believed the greatest opportunity for gains lay.

8. As noted in the previous chapter, early bond elections were contentious because of political allegations of land speculation as described in the movie *Chinatown*. See also Hoffman (1981, 141–54); Kahrl (1982, 90–103, 195); Ostrom (1953, 58, 149–54); and Nadeau (1950, 29–41). Contemporary water financing issues are discussed in Smith (2001).

9. Ostrom (1953, 50, 63).

10. Daily flows in the aqueduct from 1920 to 1935 are in cubic feet per second, from Ostrom (1953, 22). These are converted to acre-feet per year by multiplying by 1.98, the conversion factor from cubic feet per second to acre-feet, and then by 365 to express the flow as an annual amount. The 1927 aqueduct flow was 265,231 a.f.

11. Letter from Board of Public Service Commission to landowners C. P. Crowell and S. F. Zombro, Sale of Lands file, Tape GX0004, LADWP Archives.

12. Testimony from Mrs. G. L. Wallace, Transcript of Proceedings, August 13, 1926, Ladies Committee, to Board of Water and Power Commissioners, Owens River and Big Pine Canal file, Tape GX0003, LADWP Archives.

13. Resolution, July 20, 1925, Special Owens Valley Committee file, Tape GX0004, Board of Water and Power Commissioners, LADWP Archives.

14. Memo, July 21, 1926, Special Owens Valley Committee File, Board of Water and Public Service Commissioners, Tape GX0004, LADWP Archives.

15. Letter from E. F. Leahey to the Owens Valley Appraisal Committee, September 10, 1926, Owens River and Big Pine Canal file, Tape GX0003, LADWP Archives.

16. Between 1920 and 1935, Inyo County (Owens Valley) had an increase in farm size of 248 percent, largely due to Los Angeles's consolidation of properties. No other comparable Great Basin county in the area had a similar size increase. As noted in the text, farms elsewhere started larger.

17. Data in the Porter file includes water acre-feet per property along with designed pool membership. With this information it is possible to calculate the total water acre-feet available from the valley and accounted for by each pool group. The total was 266,429 a.f. with the pools providing 43,480 a.f.

18. Memo from E. F. Leahey, DWP Land Agent, to W. B. Mathews, DWP, July 26, 1928, E. F. Leahey file, Tape GX0002, LADWP Archives.

19. For discussion of bargaining externalities, see Segal (1999).

20. Herfindahl indexes based on water acre-feet give similar relative values.

21. "Owens River Canal Properties" and letter from various individuals to F. Del Valle, President, Los Angeles Water Board, February 24, 1926, Sale of Lands file, Tape GX0004; and "Tabulation Showing Status of Ranch Land Purchases," April 16, 1934, Porter file.

22. Memo, by Board of Water and Public Service Commissioners, July 21, 1926, Special Owens Valley Committee file, Tape GX0004, LADWP Archives; "Owens River Canal Properties," Sale of Lands file, Tape GX0004, LADWP Archives; and letter, from the Purchasing Committee to the Board of Water and Power Commissioners, July 21, 1925, Owens River and Big Pine Canal file, Tape GX0003, LADWP Archives.

23. See Goodenough's discussion of the offer and his acceptance in a letter, October 29, 1929, in Edward Goodenough papers, Eastern California Museum, Independence, Calif.

24. Letter from C. D. Carll, representing the Water Board, to Goodenough, July 17, 1929, Goodenough papers, Eastern California Museum.

25. See the reviews of the farm characteristics and Goodenough's justification for valuation in the Goodenough Papers, Eastern California Museum.

26. Examination of the Porter file for 1934 sales does not show those properties listed as not available. It is possible that the farmers listed in 1929, but missing in the final property list, had sold to third parties.

27. "Cashbaugh Pool," Fish Slough file, Tape GX0001, LADWP Archives.

28. Nadeau (1950, 104); Kahrl (1982, 308).

29. Kahrl (1982, 306–10) finds considerable conspiracy in the bank's failure, charging that the city's devastation of the local economy was an important factor. He does not mention, however, that rural banks were failing everywhere in the West as the agricultural economy deteriorated.

30. W. B. Mathews, 1925 letter outlining dispute over Longyear properties, W. B. Mathews file, Tape EJ00086, LADWP Archives.

31. Letter from LADWP to landowners Crowell and Zombro, Sale of Lands file, Tape GX0004, LADWP Archives; "Tabulation Showing Status of Ranch Land Purchases," April 16, 1934, Porter file.

32. Testimony from Mrs. Wallace, August 13, 1926.

33. Charles H. Lee and S. B. Robinson, "Use of Water for Irrigation in Owens Valley in Connection with the Supply of the Los Angeles Aqueduct," February 1910, p. 5, Ms/1 98.17, Lee Folder, Water Resources Research Center Archives, U.C. Berkeley.

34. $y_i = \gamma_0 + \gamma_1 x_1 + \gamma_2 x_i^2$ with y_i cultivated acreage for each of 525 farms and x_i water acre-feet per farm. The coefficient estimates for the water variables are .035 (.003) and $-1.39e-06$ (1.46e-07), standard errors in parenthesis.

35. Ellen Hanak reported that irrigation use in California typically ranges from 3 to 6 acre-feet/acre, giving the .33 and .17 figures reported in the text.

36. According to the 1925 Agricultural Census, the value of farm production per farm in Inyo County was $3,412.

37. Annual Los Angeles population change is estimated and provided at http://www.laep.org/target/science/population/table.html. The estimations are based on decennial census data and estimates provided by the California Taxpayers Association and the Los Angeles Chamber of Commerce, and those compiled by the Los Angeles County Regional Planning Commission. Aqueduct flow information is provided in Chapter Three, Table 3.3.

38. A duration (hazard) model estimation gives very similar results. The estimating equation is

$$y_{2i} = \gamma_0 + \gamma_1 x_{1i} + \gamma_2 x_{1i}^2 + \gamma_3 x_{2i} + \gamma_4 x_{2i}^2 + \gamma_5 x_{3i} + \gamma_6 x_{3i}^2 + \gamma_7 x_{4i}$$
$$+ \gamma_8 x_{5i} + \gamma_9 D_r + \gamma_{10} D_k + \gamma_{11} D_c + \gamma_{12} D_w + \gamma_{13} D_o + \eta_i$$

where y_{1i} is per acre sales price; y_{2i} is year of purchase; x_{1i} is cultivated acreage per farm; x_{2i} is total farm acreage; x_{3i} is water acre-feet/acre; x_{4i} is the change in Los Angeles's population from the previous year; x_{5i} is current aqueduct flows per capita; x_{6i} is the cumulative percentage of total water purchased as of each property's sale and D_r, D_k, D_c, D_w, D_o are dummy variables for having riparian water rights, membership in the Keough, Cashbaugh, and Watterson pools respectively, or owning a farm on a ditch but not being in a pool.

39. The estimating equation is

$$y_{1i} = \beta_0 + \beta_1 x_{1i} + \beta_2 x_{1i}^2 + \beta_3 x_{2i} + \beta_4 x_{2i}^2 + \beta_5 x_{3i} + \beta_6 x_{3i}^2 + \beta_7 x_{6i}$$
$$+ \beta_8 D_r + \beta_9 D_k + \beta_{10} D_c + \beta_{11} D_w + \beta_{12} D_o + \varepsilon_i$$

40. Substituting the pool dummies with the relevant Herfindahl indices in the estimation provides similar rankings for the pool results reported here.

41. Los Angeles's share of Colorado River aqueduct water and costs are from Hundley (2001, 229) and Erie (2000, 155 n. 26). Los Angeles sought 1.1 million acre-feet, although subsequent court rulings reduced this to 550,000 acre-feet.

The conversion is based on the present value of an annuity at 3 percent for forty years. Three percent is the mean high-grade municipal bond rate between 1920 and 1960, and the conversion factor for presenting the stock price as an annual flow is 23.11477.

42. Calculated from the land sales data in the Porter File, LADWP Archives.

43. "Statement to Mayor's Advisory Committee Prepared by the Special Owens Valley Committee of the Board of Public Service Commissioners," December 16, 1924, Special Owens Valley Committee file, Tape GX0004, LADWP Archives.

44. "Recent Purchases of Water in Owens Valley by City of Los Angeles," November 1923, Cope Rand Means Co., Engineers, San Francisco, Lee Folder, MS 7611 98246, Water Resources Research Center Archives, U.C. Berkeley.

45. Telegram from E. F. Leahey to W. B. Mathews and H. A. Van Norman, referencing an article in the *Inyo Independent*, August 11, 1928, to the effect that the constitutionality of the reparation law of 1925 would be tested in a case filed in the State Supreme Court by Jess Hessian, trustee for Watterson brothers, for damages totaling $227,182 to their hardware and garage business and other property value depreciation (Reparations July 1927–May 1935, Town Lots file, Tape GX0004, LADWP Archives).

46. Correspondence file for January–May 1927, Tape GX00086, LADWP Archives; and Legislation and Reparations file, Tape GX0004. *Statutes and Amendments to the Codes of California*, 1925, pp. 251–53, Chapter 109: "An Act providing for and relating to damages resulting from or caused by the acquisition of a water supply or taking, diverting and transporting of water from a watershed and from the lands, streams and waters therein, to and for use in and by municipal corporations and providing for compromise, arbitration and settlement and payment of claims for any such damage and empowering and authorizing municipal corporations furnishing water to or for use in municipal corporations to pay, compromise, arbitrate and settle any and all damage caused, or claimed to have been caused, by the acquiring of such land or water, or the taking, diverting or transporting of the same for the purposes aforesaid, fixing the time for the presentment of such claims and relating generally to such damages," approved by the governor, May 1, 1925. Section 1 makes municipal corporations liable for damages to property in watersheds from the acquisition and export of water. Section 2 makes municipal corporations liable for past damages from such actions. All claims must be filed within two years. Section 3 allows for joinder of claimants; claims could be arbitrated singly or collectively. Claimants can empower a committee to act on their behalf. Section 4 authorizes municipal corporations to enter into agreements to arbitrate disputes over

claims of damages. Section 5 allows for any section found unconstitutional not to affect the other sections.

47. Letter from W. W. Watterson, Chairman of the Committee from the Owens Valley Farming and Business Interests, to L. K. Chase, Chairman of the Power and Reclamation Committee of the Los Angeles Chamber of Commerce, May 29, 1924, Chamber of Commerce file, Tape GX0001, LADWP Archives.

48. Ostrom (1953, 123–27).

49. Hoffman (1981, 176–202). Letter to Board, May 4, 1927, Correspondence file for January–May 1927, Tape GX00086, LADWP Archives. Reparations claims were filed against Los Angeles by the Owens Valley Reparations Association for $2,273,721.74 and by the Big Pine Reparations Association for $538,404.76. In a letter dated May 6, 1927, the board recommended to the city clerk that they be denied.

50. Ostrom (1953, 126–27); and Press Release, Department of Water and Power, August 18, 1930, regarding plans to purchase properties now that court rulings gave the agency the latitude to do so (Correspondence file for July–August 1930, Tape EJ00086, LADWP Archives).

51. Owens Valley Town Lots file, Tape GX0005, LADWP Archives.

52. See demands by Senators Joe Riley and Dan Williams from Inyo County regarding their properties, and resolution by Senator Herbert Johnson Evans, March 9, 1931, Chair of the Conservation Committee (Investigating Committee file, Tape GX0002, LADWP Archives).

53. "Statement of Status of Owens Valley Ranch and Town Property Purchase Program by City of Los Angeles," Investigating Committee file, Tape GX0002, LADWP Archives.

54. A. J. Ford, E. A. Porter, C. D. Carll, "Status of Owens Valley Ranch Land and Town Purchases as of May 1, 1933," pp. 11–15, 17, in *A Resume of Authority, Methods Employed, Results Obtained, and Present Status of Land and Water Rights Purchasing Program of the Board of Water and Power Commissioners of the City of Los Angeles in Owens River Valley, California, May 1, 1933*, Right of Way and Land Division of the Department of Water and Power of the City of Los Angeles, Correspondence file for May–July 1933, Tape EJ00087, LADWP Archives.

55. Herbert Johnson Evans, Chairman, Conservation Committee, Resolution to Investigate Los Angeles' Purchase of Land in Inyo and Mono Counties, March 9, 1931, 35th Senate District, California, 49th Session, 1931, Investigating Committee file, Tape GX0002, LADWP Archives.

56. "Report of the Senate Special Investigating Committee on Water Situation in Inyo and Mono Counties," May 7, 1931, p. 5, LADWP Archives.

57. "Report of the Senate Special Investigating Committee," May 7, 1931, p. 9.

58. Schedule E, California State Board of Equalization, *Annual Report*, 1929–30; Schedule E, California State Board of Equalization, *Annual Report*, 1919–20. There was virtually no inflation in the 1920s that could have affected appraisals.

59. Telegram from E. A. Porter to A. J. Ford, February 11, 1931, Reparations file, Tape GX0004, LADWP Archives. A major case was *Hillside Water Company v. City of Los Angeles*, 10 Cal. 2nd 677, February 16, 1938.

60. Hoffman (1981, 255–57).

61. Letter from H. A. Van Norman, W. Turney Fox, and A. J. Ford to the Senate Conservation Committee, March 13, 1931, Investigating Committee file, Tape GX0002, LADWP Archives.

62. Wood (1973, 57).

63. Ostrom (1953, 126); "Owens Valley Situation, Synopsis," Correspondence file for January–March 1929, Tape EJ0086; and memo by A. J. Ford, land agent for the investigating committee, March 11, 1931, Investigating Committee file, Tape GX0002, LADWP Archives. Los Angeles paid $2,975,833 for Bishop lots, $772,635 for Big Pine lots, $730,306 for Independence lots, $102,446 for Laws lots, and $1,217,560 for Lone Pine lots, for a total of $5,798,780.

64. Nadeau (1950, 125–30).

65. Hoffman (1981, 253). Of the May 20, 1930, bond issue of $38,800,000, $19,181,000 was to buy up remaining properties, $7,400,000 to build a tunnel from Mono Basin, and the rest for dam construction in Long Valley and other places.

66. Ford, Porter, and Carll, "Status of Owens Valley Ranch Land." pp. 22–23. As discussed by Ostrom (1953, 126–27), the California Supreme Court ruled that the city could buy town properties; approval of the bond issue came December 23, 1930. During 1931 and 1932, negotiations proceeded. The board required that that each town lot owner reject further reparations claims.

67. Ostrom (1953, 121–27).

68. *Population*, Volume VI, "Families, Reports by States, Giving Statistics for Families, Dwellings and Homes, by Counties, for Urban and Rural Areas and for Urban Places of 2,500 or More," Table 45, p. 38 (Washington, D.C.: Government Printing Office, 1933).

NOTES TO CHAPTER 5

1. California State Board of Equalization, *Annual Reports:* for 1923–34, Schedule E; 1910, Schedule G; 1911–14, Schedule H; 1915–16, Schedule G; 1917–18, Schedule F; 1919–22, Schedule D; 1923–34, Schedule E. There was virtually no inflation in the 1920s that could have affected appraisals.

2. Pisani (1984, 303).

3. For example, the California State Board of Equalization, *Annual Report, 1900*, Schedule D; and *Annual Report, 1930*, Schedule C.

4. Most recently, see *New York Times*, August 8, 2004, p. 14, where the Owens Valley transfer is labeled a "Century-Old Land Grab."

5. For example, according to the U.S. Agricultural Census, 1910, per acre land and building values were $52 in Inyo, $24 in Lassen, $27 in Churchill, $36 in Douglas, and $32 in Lyon County. By 1930 the relative values were $143, $21, $80, $40, and $41. In 1930, per farm values were $62,200 for Inyo, $21,109 for Lassen, $11,565 for Churchill, $44,658 for Douglas, and $21,891 for Lyon.

6. Barnard and Jones (1987, 10–12).

7. Census data are from http://fisher.lib.virginia.edu/cgi-local/censusbin/census/cen.pl. These figures are only representative of the actual gains from trade. The data for Los Angeles do not include increases in urban land values, and the amount of agricultural land in Los Angeles declined by 369,000 acres between 1900 and 1930. Similarly, the amount of farmland in Inyo County declined by 46,000 acres, whereas in Lassen County, farmland grew by 92,000 acres. Nevertheless, the data are indicative of the values involved.

8. These values are calculated as follows: Using census data for 1900 and 1930, Inyo County had 141,059 acres in farms in 1900 and Lassen 381,103 acres. In 1930, Lassen had 473,268 acres, an increase of 24 percent. Had Inyo farm acreage grown in the same way, then in 1930 there would have been 174,913 farm acres. Lassen farm acreage values doubled over the thirty years, and 1900 Inyo per acre values were $13. Using the Lassen increase gives a 1930 per acre value in 1930 of $26, and multiplying this times the 1930 estimated acreage gives a value of farm acreage of $4,547,738. Actual census value of farmland and buildings in 1930 was $13,559,534, for a difference of $9,011,796.

9. When the gains from trade are very large, distributional outcomes move to the forefront, as they did in Owens Valley negotiations. Usually, trades are smoother when the benefits are shared reasonably equally, but encounter more difficulties in completion when the distribution is very skewed toward one party. My colleague P. J. Hill suggested this point to me.

10. This figure is based on the difference in the rise in value of agricultural land and buildings in Los Angeles County and Inyo County between 1900 and 1930—$407,051,000 as compared to $11,568,000 (U.S. Census data). On a per capita basis the distribution is different, but Owens Valley farmers were after the aggregate.

11. Ellen Hanak of the Public Policy Institute of California reminded me of the nature of supply and demand forces in generating this result.

12. Dellapenna (2000, 356–57) also points to the importance of distributional concerns in water markets. For additional discussion, see Libecap (2005).

NOTES TO CHAPTER 6

1. Hundley (2001, 336–62) is a notable exception.

2. Frequently quoted. See for example, Hundley (2001, 155).

3. "Chronological Statement of Some Facts Pertaining to Land, Construction, Water Supply and Organization Matters of the Department of Water and Power in the Owens Valley District from December 10, 1928, to February 1, 1945, including a General Statement of Facts from 1895 to December 9, 1928," February 1, 1945, Bibliography file, Tape EJ00087, LADWP Archives.

4. Los Angeles Department of Water and Power (LADWP) (1976, 1-1, 1-2).

5. LADWP (1976, 5–13).

6. Kahrl (1982, 352).

7. LADWP (1976, 2-2). The United States Geological Survey (USGS) reported 2.48 acre-feet per acre as the average for U.S. agriculture. See http://ga.water .usgs.gov/edu/wuir.html.

8. "Chronological Statement of Some Facts," February 1, 1945; LADWP (1976, 2-2).

9. LADWP (1976, 5-12).

10. Walton (1992, 201); Kahrl (1982, 344).

11. Kahrl (1982, 401). This area had long been contested and was part of the Hillside Decree regarding water rights and groundwater pumping. See Galloway et al. (2003, 14).

12. http://wsoweb.ladwp.com/Aqueduct/historyoflaa/aqueductfacts.htm. There are various capacity figures for the first aqueduct, depending on the time given. The aqueduct's capacity was gradually increased with side extensions and other improvements.

13. LADWP (1976, 2-4); Groeneveld (1992, 136).

14. LADWP (1976, 2-4, 2-5, 5-11, IV-10).

15. Groeneveld (1992, 137). Galloway et al. (2003) described damages to plants from increased groundwater pumping, arguing that by 1981 native plants dependent on shallow groundwater sources were reduced by 20–100 percent on about 26,000 acres.

16. Kahrl (1982, 415, 416); Walton (1992, 246–47, 249).

17. 1433 Statutes of 1970.

18. Kahrl (1982, 416).

19. Kahrl (1982, 418).

20. Kahrl (1982, 419).

21. Kahrl (1982, 420–22); Walton (1992, 252).

22. Walton (1992, 253–54, 258).

23. Kahrl (1982, 397).

24. Kahrl (1982, 387).

25. Walton (1992, 260–61).

26. Kahrl (1982, 424–28).

27. http://www.inyowater.org/Water_Resources/icwaterpolicy.html.

28. Walton (1992, 270); Galloway et al. (2003).

29. http://www.inyowater.org/Water_Resources/water_agreement/default. html; Groeneveld (1992, 141).

30. www.inyowater.org/Annual_Reports/2002-2003/Water_Agreement _status. Saltcedar is an invasive species that was introduced into the western United States in the nineteenth century. It consumes a great deal of water and crowds out native plant species.

31. "Agreement Between the County of Inyo and the City of Los Angeles and Its Department of Water and Power on a Long Term Ground Water Management Plan for Owens Valley and Inyo County," Section XII (http://www.inyo water.org/Water_Resources/water_agreement/default.html).

32. Los Angeles Department of Water and Power (LADWP), Environmental Protection Agency (EPA), and Inyo County (2002).

33. Superior Court of Inyo County, SICVCV-01-29768, February 13, 2004.

34. http://www.ladwp.com/ladwp/cms/ladwp005749.jsp.

35. Superior Court of Inyo County, SICVCV-01-29768, July 25, 2005.

36. LADWP (2004, 2-4 through 2-7). The agency used 3 percent over fifteen years in its present-value calculations.

37. As outlined in the Final EIR/EIS, LADWP (2004, 2-5 through 2-7), there was to be cost sharing for the non–pump station operating costs between Los Angeles and Inyo County, but the LADWP provides matching funds for saltcedar control and other expenses and provides most of the labor time. There also was EPA funding, largely to Inyo County.

38. LADWP (2004, 2-15).

39. LADWP (2004, 2-33, 2-39).

40. LADWP, Environmental Protection Agency (EPA), and Inyo County (2002, 10-8 to 10-14). The steady-state figure is likely too low since the 2002 EIS/ EIR assumed that a 150 cfs pump station would capture much of the delta bypass flow. But that feature was dropped for the 2004 EIS/EIR.

41. Those figures were fifteen years at 3 percent. 2005 Urban Water Management Plan, p. 3-3 (http://www.ladwp.com/ladwp/cms/ladwp007157.pdf).

42. See Cahill et al. (1996).

43. Schade (2002); Gill and Cahill (1992, 66).

44. Gill and Cahill (1992, 63).

45. http://www.gbuapcd.org/OVPM10SIP.htm.

46. Dust control issues were arising also as part of the Mono Lake controversy. See Hart (1996, 104).

47. LADWP (2004, 12-17).

48. GBUAPCD, 2003, State Implementation Plan, Summary, p. S-7 (http://www.gbuapcd.org/ovpm10sip.htm).

49. GBUAPCD, 2003, State Implementation Plan, Chapter 7, p. 7.8 (http://www.gbuapcd.org/ovpm10sip.htm). The GBUAPCD used 5 percent and twenty-five years for the capital cost estimations. Three percent and fifteen years are used in the text to be comparable with other present-value calculations presented in the chapter.

50. Water cost estimate provided by Richard Harasick, director of the LADWP's Owens Lake Dust Mitigation Project, personal communication, March 31, 2005.

51. 2005 Water Management Plan, pp. 3-6. The price used is the average of tier 1 and tier 2 water for $501/acre-foot (http://www.ladwp.com/ladwp/cms/ladwp007157.pdf).

52. See discussion of the health impact of PM-10 pollution, Final EIR/EIS, 3.3a–3.4, Table 3.3; Gill and Cahill (1992, 67). See also the weak evidence of any health effect in Sarah Kittle, "Survey of Reported Health Effects of Owens Lake Particulate Matter," January 14, 2000 (http://www.gbuapcd.org/information/OwensLake Particulate Matter Health Effects.htm).

53. California Department of Health Services (2002); California Department of Health Services (2000a); American Lung Association (2005).

54. Gill (1995, 145–46).

NOTES TO CHAPTER 7

1. Wallace C. McPherson file, Tape GX0003, LADWP Archives.

2. "Status of Owens Valley Ranch Land and Town Purchases as of May 1, 1933," Correspondence file for May–July 1933, Tape EJ00087, LADWP Archives.

3. *Los Angeles v. Aiken*, Cunningham Property file, *Inyo Independent*, October 2, 1933, Tape GX0008, LADWP Archives.

4. Jones and Stokes Associates (1993, 3G-5).

5. "Points and Authorities in Support of Legal Opinion in Securing Water in Mono Basin for Use in Los Angeles," Correspondence file for July 1930, Tape EJ00086, LADWP Archives.

6. Letter to Board of Water and Power Commissioners from Erwin Werner, City Attorney, December 8, 1930. See also letter from W. B. Mathews, Special Counsel, to Board of Water and Power Commissioners, March 22, 1930, regarding the rights of riparian water rights holders to undiminished flows on Rush, Parker, Walker, and Lee Vining Creeks and Mono Lake. He stated that appro-

priators have no right against riparian owners. Both letters are in the Mono Basin file, Tape GX0003, LADWP Archives.

7. See *Los Angeles v. Aiken*, No. 2449, Summons and Complaint, December 1931–April 1934 file, Tape GX0008, LADWP Archives. The complaint lists the targeted property owners as defendants, states that the creeks are nonnavigable streams in Mono County, lists their maximum flows, and asserts, "Plaintiffs have the right to take, divert and use all of the waters flowing or to flow in each of said streams at and below the points hereinafter specified, as against the defendants and each of them, and against all the world, except as to the rights of the defendants herein which are sought to be condemned in this action." This case started in Mono County Superior Court and continued through 1934 in the Tuolumne County Superior Court.

8. See letter from W. B. Mathews, Special Counsel, to Board of Water and Power Commissioners, March 25, 1930, Correspondence file for January–March 1930, Tape EJ00086, LADWP Archives.

9. "Statement of Fact Respecting Matters Involved in Purchase of Properties and Water Rights Incidental to the Carrying Out of the Mono Basin Project," June 1933, Jon. W. Baumgartner, President of the Board; H. A. Van Norman, Chief Engineer; A. F. Southwick, Member of the Board; E. F. Scattergood, Chief Electrical Engineer and General Manager; Mark L. Herron, Assistant City Attorney; T. B. Cosgrove, Special Counsel, Department of Water and Power. This was the Mono Basin Committee to investigate Southern Sierras Power properties and others. Tape GX0002, Land and Water Rights File, LADWP Archives.

10. Hart (1996, 38–40).

11. "Chronological Statement of Some Facts Pertaining to Land, Construction, Water Supply and Organization Matters of the Department of Water and Power in the Owens Valley District from December 10, 1928, to February 1, 1945, Including a General Statement of Facts from 1895 to December 9, 1928," February 1, 1945, Bibliography file, Tape EJ00087, LADWP Archives.

12. 2005 Urban Water Management Plan, p. 3-1, LADWP; http://www.ladwp.com/ladwp/cms/ladwp007157.pdf; Gaines (1981, 66).

13. http://www.mmmfiles.com/chap20.htm.

14. Hart (1996, 38–40).

15. Mono Basin Clearinghouse (www.monobasinresearch.org/timelines/polchr.htm), Appendix R, Mono Basin EIR, May 1993, Legal History.

16. Many western states, including California, now allow for instream flows and salvaged water from conservation to be consistent with beneficial use. The "growing communities" doctrine also now helps to protect municipal water agencies from loss due to nonuse. See Carpenter (1997).

17. "Chronological Statement of Some Facts," February 1, 1945.

18. "Chronological Statement of Land, Construction and Organization Matters in the Owens Valley District" from 1896 to 1945 by E. A. Porter, Miscellaneous file, Tape EJ00087, LADWP Archives.

19. Hart (1996, 118). See also, Hundley (2001, 336–46) for discussion of the politics of the Mono Lake battle.

20. Hart (1996, 56).

21. Kahrl (1982, 405–6); Hart (1996, 56).

22. http://wsoweb.ladwp.com/Aqueduct/historyoflaa/aqueductfacts.htm.

23. *National Audubon v. Superior Court*, 33 Cal. 3d 429.

24. Jones and Stokes Associates (1993, S-1); Dunning (1990, 20); Hart (1996, 56–58).

25. Calculated from sources for Figure 7.2.

26. Kahrl (1982, 429–30).

27. Botkin et al. (1988, ix).

28. Gaines (1981, 14); Hart (1996, 60).

29. Gaines (1981, 87–88); Gill (1995).

30. Kahrl (1982, 431–32).

31. Gaines (1981, 95–99); Hart (1996, 84–88); and Kahrl (1982, 431–34). Kahrl is not at all sympathetic to Los Angeles's predicament, suggesting that alternatives and conservation could offset the lost water, a theme carried by many proponents of restrictions on Mono Basin water exports.

32. For a classic discussion of the Public Trust Doctrine, see Sax (1970).

33. Hart (1996, 98).

34. In this regard, see Koehler (1995); Blumm and Schwartz (1995); and Arnold (2004).

35. Conway (1984, 634 n. 108).

36. *City of Los Angeles Department of Water and Power v. National Audubon Society et al.*, No. 83-300, 464 U.S. 977, November 7, 1983.

37. Hart (1996, 103).

38. Hart (1996, 109–14).

39. Hart (1996, 115–17).

40. *The Future of Mono Lake*, CORI, Community and Organization Research Institute (UCSB), as described by Hart (1996, 124–25).

41. Hart (1996, 132).

42. Jones and Stokes Associates (1993, 3D-113); Hart (1996, 139–40).

43. Mono Basin Clearinghouse (www.monobasinresearch.org/timelines/polchr.htm).

44. Hart (1996, 148).

45. Jones and Stokes Associates (1993, 3H-6, 8, 30).

46. Hart (1996, 144).

47. Hart (1996, 162).

48. Mono Basin Clearinghouse, Mono Lake Committee (www.monobasin research.org/onlinereports/eirprofile.htm).

49. 2005 Urban Water Management Plan, City of Los Angeles, pp. 3-5 to 3-7, DWP (http://www.ladwp.com/ladwp/cms/ladwp007157.pdf).

50. Jones and Stokes Associates (1993, 3J-19, 19, Table 3N-19). Alternative water would require higher treatment costs; power generation would fall by 134,000 MWh/year. This is power that does not generate greenhouse gases.

NOTES TO CHAPTER 8

1. *National Audubon Society v. Superior Court*, 33 Cal 3d 419. See also Dunning (1990, 19); Kahrl (1982, 430–44, 718–23); and Mono Basin Clearinghouse (www.monobasinresearch.org/timelines/polchr.htm); and Jones and Stokes Associates (1993, Appendix R, "Legal History").

2. See Meyers (1989) and Stevens (2003) for expansion to rivers and wildlife.

3. Arnold (2004, 49).

4. As an example of the continuing, unresolved nature of the conflict in Owens Valley, see Jeffrey Anderson, "The Eternal Dustbowl; Paying for the sins of L.A.'s water barons has created a half-billion-dollar boondoggle," *L.A. Weekly*, March 22, 2006.

5. 2005 Urban Water Management Plan, City of Los Angeles, p. ES-8, DWP (http://www.ladwp.com/ladwp/cms/ladwp007157.pdf). Using $501 as the price of alternative water and discounting for twenty-five years at 5 percent. Mean price of tier 1 and tier 2 MWD water.

6. In the Final EIR for the Lower Owens River Project, the LADWP presented other cost projections in LADWP (2004, Section 12, p. 18): "As a result of these projects [reduced Mono Basin exports, reduced Owens Valley groundwater pumping, and the dust control project], the total cumulative reduction from the amount exported via the Los Angeles Aqueduct could be as high as a total of 131,603 acre-feet per year. In order for LADWP to replace the water that is not exported from the Owens Valley and the Mono Basin, it will have to purchase water from Metropolitan Water District (MWD) for the foreseeable future at a cost of $350 per acre-foot or $46.1 million dollars annually. This will create additional demand and impacts on MWD supplies that are already being impacted by other significant water related issues."

7. Kahrl (1982, 444, 448); Hart (1996, 65, 182); Walton (1992, 246); and Gaines (1981).

8. See http://www.owt.org/. Libecap (2005, 19–23) describes some of the transaction costs of such exchanges, including bilateral monopoly, valuation, and third-party effects.

9. The uses and problems of eminent domain and just compensation are outlined in Fischel (1998). Eminent domain has been used to acquire private in-holdings in national parks.

10. See Jones and Stokes Associates (1993, Chapter 3N, Appendix X). Also see Loomis (1987).

11. There is, of course, potential for conflict in these compulsory exchanges, but since compensation was provided, as compared to little or none under the public trust doctrine, it seems likely that there would have been less contention.

12. See Barzel (1997) and Libecap (1989).

13. See Hart (1996, 3, 181) and Dunning (1990, 21). For positive assessments of the public trust doctrine's potential for environmental regulation, see Arnold (2004) and Blumm and Schwartz (1995).

14. Ostrom (1990) discusses when common property is effective.

15. See Libecap (2006) for discussion. See also Stavins (2003) for discussion of the movement toward market-based instruments and tradable emission permits; Hannesson (2004) discusses ITQs; and Libecap and Smith (1999) oil-field unitization.

16. Stavins (2003); Libecap and Smith (1999).

17. *Marks v. Whitney,* 6 Cal. 3d 251 (1971); *Baker v. Mack,* 19 Cal. App. 3d 1040 (1971); *National Audubon Society v. Superior Court,* 33 Cal. 3d 425, 434–37.

18. *Audubon Society v. Superior Court,* 33 Cal. 3d 444.

19. *Audubon Society v. Superior Court,* 33 Cal. 3d 440.

20. For analysis of legal disputes and their resolution, see Cooter and Rubinfeld (1989).

21. For discussion, see Goodman (1978, 394) and Katz (1988, 129, 136).

22. The problem of unequal costs among litigants and the impact on judicial rulings is discussed by Katz (1988, 129).

23. Graduate students and other volunteers gathered data on the Mono Lake ecosystem for the Mono Lake Committee. See Hart (1996, 65, 182); Walton (1992, 246); and Gaines (1981), for example. Hart's (1996) excellent account of conflict over Mono Basin water is generally supportive of the role of passionate advocates of stopping Los Angeles's diversions and their efforts through the courts and the regulatory process. Similarly, see Arnold's (2004, 18) discussion of the campaign to "Save Mono Lake."

24. This lesson is in contrast to the one drawn by Arnold (2004, 50), for example.

REFERENCES

Allen, Douglas W. 2000. "Transaction Costs." In *Encyclopedia of Law and Economics*, ed. Boudewijn Bouckaert and Gerrit De Geest, Vol. 1: 893–926. Cheltenham, UK: Edward Elgar.

American Lung Association, Epidemiological and Statistical Unit, Research and Scientific Affairs. 2005. "Estimated Prevalence and Incidence of Lung Disease by Lung Association Territory." http://www.lungum.org/ia/pdf/EstPrev05.pdf.

Anderson, Terry, and Ronald N. Johnson. 1986. "The Problem of Instream Flows." *Economic Inquiry* 24 (4): 535–53.

Arnold, Craig Anthony. 2004. "Working Out an Environmental Ethic: Anniversary Lessons from Mono Lake." *Wyoming Law Review* 4: 1–55.

Barnard, Charles, and John Jones. 1987. *Farm Real Estate Values in the United States By Counties, 1850–1982*, Washington, D.C.: USDA, ERS, Government Printing Office.

Barzel, Yoram. 1982. "Measurement Cost and the Organization of Markets." *Journal of Law and Economics* 25 (1): 27–48.

——— 1997. *Economic Analysis of Property Rights*. 2d ed. New York: Cambridge University Press.

Bateman, Paul, Dorothy Cragen, Mary DeDecker, Raymond Hock, E. P. Pister, Paul Lane, Antonio Rossman. 1978. *Deepest Valley: A Guide to Owens Valley, Its Roadsides and Mountain Trails*. Edited by Genny Schumacher Smith. Revised edition. Los Altos, Calif.: W. Kaufmann.

Beck, Robert E. 2000. "The Regulated Riparian Model Water Code: Blueprint for Twenty First Century Water Management." *William and Mary Environmental Law and Policy Review* 25 (Fall): 113–67.

Blair, Rodger D., David L. Kaserman, and Richard F. Romano. 1989. "A Pedagogical Treatment of Bilateral Monopoly." *Southern Economic Journal* 55 (4): 831–41.

Blumm, Michael C., and Thea Schwartz. 1995. "Mono Lake and the Evolving Public Trust in Western Water." *Arizona Law Review* 37: 701–38.

Bokum, Consuelo. 1996. "Implementing the Public Welfare Requirement in New Mexico's Water Code." *Natural Resources Journal* 36 (Fall): 681–713.

Botkin, Daniel B., Wallace S. Broecker, Lorne G. Everett, Joseph Shapiro, and John A. Wiens. 1988. *The Future of Mono Lake.* Report of the Community and Organization Research Institute "Blue Ribbon Panel" for the Legislature of the State of California. Report no. 68. Riverside, Calif.: Water Resources Center, University of California.

Brown, Thomas C. 2006. "Trends in Water Market Activity and Price in the Western United States." *Water Resources Research* 42 (9): W09402.

Cahill, Thomas A., Elizabeth A. Gearhart, Thomas E. Gill, Dale A. Gillette, and Jeffrey S. Reid. 1996. "Saltating Particles, Playa Crusts and Dust Aerosols at Owens (dry) Lake, California." *Earth Surface Processes and Landforms* 21 (July): 621–39.

California Department of Health Services, Environmental Health Investigations Branch. 2000a. "California County Asthma Hospitalization Chart Book: Data for 1995–1997." http://www.dhs.ca.gov/ehib/EHIB2/PDF/Hosp %20Chart%20Book%202000.pdf.

———— 2000b. "California County Asthma Mortality Chart Book: Data for 1990–1997." http://www.dhs.ca.gov/ehib/EHIB2/PDF/Mortality%20Chart %20Book.pdf.

———— 2002. "Chronic Lower Respiratory Disease Deaths, California 2002." http://www.dhs.ca.gov/ehib/EHIB2/PDF/Hosp%20Chart%20Book%2020 .pdf.

Carey, Janis M., and David L. Sunding. 2001. "Emerging Markets in Water: A Comparative Analysis of the Central Valley and Colorado Big Thompson Projects." *Natural Resources Journal* 41 (Spring): 283–328.

Carpenter, Janis E. 1997. "Water for Growing Communities: Refining Tradition in the Pacific Northwest." *Environmental Law* 27: 127–49.

Carter, Harold O., Henry J. Vaux Jr., and Ann F. Scheuring, eds. 1994. *Sharing Scarcity: Gainers and Losers in Water Marketing.* Davis: University of California Agricultural Issues Center.

Coase, Ronald H. 1937. "The Nature of the Firm." *Economica* 4: 386–405.

———— 1960. "The Problem of Social Cost." *Journal of Law and Economics* 3: 1–44.

Cohen, Jane Maslow. 2005. "Symposium: Of Waterbanks, Piggybanks, and Bankruptcy: Changing Directions in Water Law." *Texas Law Review* 83 (7): 1809–72.

Colby, Bonnie G. 1990. "Transaction Costs and Efficiency in Western Water Allocation." *American Journal of Agricultural Economics* (December): 1184–92.

——— 1995. "Water Re-allocation and Valuation: Voluntary and Involuntary Transfers in the Western United States." In *Water Law: Trends, Policies, and Practice*, ed. Kathleen Marion Carr and James D. Crammond, 112–26. Chicago: American Bar Association.

Colby, Bonnie G., Mark A. McGinnis, and Ken Rait. 1989. "Procedural Aspects of State Water Law: Transferring Water Rights in the Western States." *Arizona Law Review* 31 (4): 697–720.

Conkling, Harold. 1921. *Report on Owens Valley Project California*. Washington, D.C.: U.S. Department of the Interior, United States Reclamation Service.

Conway, Timothy J. 1984. "Note: *National Audubon Society v. Superior Court*: The Expanding Public Trust Doctrine." *Environmental Law* 14: 617–40.

Cooter, Robert D., and Daniel L. Rubinfeld. 1989. "Economic Analysis of Legal Disputes and Their Resolution." *Journal of Economic Literature* 27 (3): 1067–97.

Dahlman, Carl J. 1979. "The Problem of Externality." *Journal of Law and Economics* 22: 141–62.

Davies, David Howard. 1999. "Great Lakes Commentary: Water Diversion from the Great Lakes." *Toledo Journal of Great Lakes Law, Science, and Policy*: 109–23.

Delameter, Charles Ellis. 1977. "The Owens Valley, City of Los Angeles Water Controversy: An Oral History Examination of the Events of the 1920's and the 1970's." Master's thesis, California State University Fullerton.

Dellapenna, Joseph W. 2000. "The Importance of Getting Names Right: The Myth of Markets for Water. *William and Mary Environmental Law and Policy Review* 25: 317–77.

Demsetz, Harold. 1964. "The Exchange and Enforcement of Property Rights." *Journal of Law and Economics* 7: 11–26.

——— 1968. "The Cost of Transacting." *Quarterly Journal of Economics* 82: 33–53.

Dunning, Harrison C. 1990. "Dam Fights and Water Policy in California: 1969–1989." *Journal of the West* 29 (3): 14–27.

Eggertsson, Thrainn. 1990. *Economic Behavior and Institutions*. Cambridge: Cambridge University Press.

Ellickson, Robert. 1993. "Property in Land." *Yale Law Journal* 102: 1315–1400.

Epstein, Richard. 1979. "Possession as the Root of Title." *Georgia Law Review* 13: 1221.

Erie, Steven P. 2000. "'Mulholland's Gifts': Further Reflections upon Southern California Water Subsidies and Growth." *California Western Law Review* 37: 147–60.

Ethier, Wilfred. 1982. "National and International Returns to Scale in the Modern Theory of International Trade." *American Economic Review* 72: 389–405.

Ewan, Rebecca Fish. 2000. *A Land Between: Owens Valley, California.* Baltimore: Johns Hopkins University Press.

Feldman, Michael P., and David B. Audretsch. 1999. "Innovation in Cities: Science-based Diversity, Specialization and Localized Competition." *European Economic Review* 43: 409–29.

Ferrie, Joseph P. 1994. "The Wealth Accumulation of Antebellum European Immigrants to the U.S., 1840–60." *Journal of Economic History* 54 (1): 1–33.

Fischel, William A. 1998. "Eminent Domain and Just Compensation." In *The New Palgrave Dictionary of Economics and the Law*, ed. Peter Newman, Vol. 2: 34–43. London: Macmillan.

Gaines, David. 1981. *Mono Lake Guide Book.* Lee Vining, Calif.: Kutsavi Books.

Galenson, David, and Clayne Pope. 1989. "Economic and Geographic Mobility on the Farming Frontier: Evidence from Appanoose County, Iowa, 1850–1870." *Journal of Economic History* 49 (3): 635–55.

Galloway, Devin L., William M. Alley, Paul M. Barlow, Thomas E. Reilly, and Patrick Tucci. 2003. "Evolving Issues and Practices in Managing Groundwater Resources: Case Studies on the Role of Science." Circular 1247. Reston, Va.: USDI, USGS.

Getches, David H. 1993. "From Askhabad, to Wellton-Mohawk, to Los Angeles: The Drought in Water Policy." *University of Colorado Law Review* 64 (1993): 523–53.

——— 1997. *Water Law in a Nut Shell.* St. Paul, Minn.: West Publishing.

Gill, Thomas Edward. 1995. "Dust Generation Resulting from Desiccation of Playa Systems: Studies of Mono and Owens Lakes, California." PhD diss., University of California, Davis.

Gill, Thomas E., and Thomas A. Cahill. 1992. "Playa Generated Dust Storms from Owens Lake." In *The History of Water: Eastern Sierra Nevada, Owens Valley, White-Inyo Mountains*, ed. Clarence Hall Jr., Victoria Doyle-Jones, and Barbara Widawski, Vol. 4: 63–73. White Mountain Research Station Symposium. Berkeley: University of California Press.

Glaeser, Edward L., Matthew E. Kahn, and Jordan Rappaport. 2000. "Why Do the Poor Live in Cities?" NBER Working Paper 7636. Cambridge, Mass.: National Bureau of Economic Research.

Glaeser, Edward L., Hedi Kallal, Jose A. Schenkman, and Andrei Shleifer. 1992. "Growth in Cities." *Journal of Political Economy* 100 (6): 1126–52.

Glaeser, Edward L., Jed Kolko, and Albert Saiz. 2001. "Consumer City." *Journal of Economic Geography* 1: 27–50.

Glennon, Robert Jerome. 1991. "'Because That's Where the Water Is': Retiring Current Water Uses to Achieve the Safe-Yield Objective of the Arizona Groundwater Management Act." *Arizona Law Review* 33: 89–114.

——— 2002. *Water Follies: Groundwater Pumping and the Fate of America's Fresh Waters.* Washington, D.C.: Island Press.

——— 2005. "Water Scarcity, Marketing, and Privatization." *Texas Law Review* 83 (7): 1873–1902.

Glennon, Robert Jerome, Alan Ker, and Gary D. Libecap. 2007. *Western Water Data Set.* Bren School of Environmental Science and Management, University of California, Santa Barbara.

Goin, Peter. 1984. "Review of *At Mono Lake.*" *San Francisco Chronicle*, January 15.

Goodman, John C. 1978. "An Economic Theory of the Evolution of the Common Law." *Journal of Legal Studies* 7 (2): 393–406.

Gould, George A. 1995. "Recent Developments in the Transfer of Water Rights." In *Water Law: Trends, Policies, and Practice*, ed. Kathleen Marion Carr and James D. Crammond, 93–103. Chicago: American Bar Association.

Gray, Brian E. 1994. "The Modern Era in California Water Law." *Hastings Law Journal* 45 (January): 249–308.

Gray, Brian E., Bruce C. Driver, and Richard W. Wahl. 1991. "Economic Incentives for Environmental Protection: Transfers of Federal Reclamation Water: A Case Study of California's San Joaquin Valley." *Environmental Law* 21 (Spring): 911–83.

Great Basin Unified Air Pollution Control District (GBUAPCD). 2003. "Owens Valley PM_{10} Planning Area Demonstration of Attainment State Implementation Plan 2003 Revision." Bishop, Calif.

Griffin, Ronald C., and Fred O. Boadu. 1992. "Water Marketing in Texas: Opportunities for Reform." *Natural Resources Journal* 32: 265–88.

Groeneveld, David P. 1992. "Owens Valley California, Plant Ecology: Effects from Export Groundwater Pumping and Measures to Conserve the Local Environment." In *The History of Water: Eastern Sierra Nevada, Owens Valley, White-Inyo Mountains*, ed. Clarence A. Hall Jr., Victoria Doyle-Jones, and Barbara Widawski, Vol. 4: 128–55. White Mountain Research Station Symposium. Berkeley: University of California Press.

Haddad, Brent M. 2000. *Rivers of Gold: Designing Markets to Allocate Water in California*. Washington, D.C.: Island Press.

Haddock, David D., and Fred S. McChesney. 1991. "Bargaining Costs, Bargaining Benefits, and Compulsory Nonbargaining Rules." *Journal of Law, Economics, and Organization* 7 (1): 334–54.

Hall, Clarence A. Jr., Victoria Doyle-Jones, and Barbara Widawski, eds. 1992. *The History of Water: Eastern Sierra Nevada, Owens Valley, White-Inyo Mountains*. Los Angeles: University of California White Mountain Research Station.

Hanak, Ellen. 2003. *Who Should Be Allowed to Sell Water in California? Third-Party Issues and the Water Market*. San Francisco: Public Policy Institute of California.

Hanak, Ellen, and Caitlin Dyckman. 2003. "Counties Wresting Control: Local Responses to California's Statewide Water Market." *University of Denver Water Law Review* 6 (Spring): 490–518.

Hanemann, W. Michael. 2005. "The Economic Conception of Water." In *Water Crisis: Myth or Reality?* ed. P. Rogers and R. Llamas. London: Taylor and Francis.

Hannesson, Rögnvaldur. 2004. *The Privatization of the Oceans*. Cambridge, Mass.: MIT Press.

Hardin, Garrett. 1968. "The Tragedy of the Commons." *Science* 162: 1243–48.

Hart, John. 1996. *Storm over Mono: The Mono Lake Battle and the California Water Future*. Berkeley: University of California Press.

Helseley, R., and W. Strange. 1990. "Matching and Agglomeration Economies in a System of Cities." *Regional Science and Urban Economics* 20: 189–212.

Henderson, J. Vernon. 1974. "The Sizes and Types of Cities." *American Economic Review* 64: 640–56.

Hoffman, Abraham. 1981. *Vision or Villainy: Origins of the Owens Valley–Los Angeles Water Controversy*. College Station: Texas A&M University Press.

Hoffman, Elizabeth, and Gary D. Libecap. 1991. "Institutional Choice and the Development of U.S. Agricultural Policies in the 1920s." *Journal of Economic History* 51 (2): 397–411.

Howitt, Richard E. 1994. "Effects of Water Marketing on the Farm Economy." In *Sharing Scarcity: Gainers and Losers in Water Marketing*, ed. Harold O. Carter et al., 97–132. Davis, Calif.: Agricultural Issues Center.

Howitt, Richard E., and Kristiana Hansen. 2005. "The Evolving Western Water Markets." *Choices* 20 (1): 59–63.

Hundley, Norris Jr. 2001. *The Great Thirst, Californians and Water: A History.* Revised edition. Berkeley: University of California Press.

Israelsen, Orson W., J. Howard Maughan, and George P. South. 1946. "Irrigation Companies in Utah: Their Activities and Needs." *Utah State Agricultural Experiment Station Bulletin* 322. Logan: Utah State Agricultural College.

Jacobs, Jane. 1984. *Cities and the Wealth of Nations: Principles of Economic Life.* New York: Random House.

Johnson, Ronald N., Micha Gisser, and Michael Werner. 1981. "The Definition of a Surface Water Right and Transferability." *Journal of Law and Economics* 24 (2): 273–88.

Johnson, Stephen, ed. 1983. *At Mono Lake: A Photographic Exhibition, Selected by Stephen Johnson, Al Weber, and Don Worth.* San Francisco: Friends of the Earth Foundation.

Jones and Stokes Associates. 1993. "Draft Environmental Impact Report for the Review of the Mono Basin Water Rights of the City of Los Angeles." Sacramento: California State Water Resources Control Board, Division of Water Rights.

———— 1994. "Final Environmental Impact Report for the Review of Mono Basin Water Rights of the City of Los Angeles." Sacramento: California State Water Resources Control Board, Division of Water Rights.

Kahrl, William L. 1982. *Water and Power: The Conflict over Los Angeles' Water Supply in the Owens Valley.* Berkeley: University of California Press.

———— 2000a. "Part I: The Politics of California Water: Owens Valley and the Los Angeles Aqueduct, 1900–1927." *Hastings West-Northwest Journal of Environmental Law and Policy* 6: 239–50.

———— 2000b. "Part II: The Politics of California Water: Owens Valley and the Los Angeles Aqueduct, 1900–1927." *Hastings West-Northwest Journal of Environmental Law and Policy* 6: 255–67.

Kanazawa, Mark. 2003. "Origins of Common-Law Restrictions on Water Transfers: Groundwater Law in Nineteenth-Century California." *Journal of Legal Studies* 32 (1): 153–80.

Katz, Avery. 1988. "Judicial Decisionmaking and Litigation Expenditure." *International Review of Law and Economics* 8: 127–43.

Kennan, John, and Robert B. Wilson. 1993. "Bargaining with Private Information." *Journal of Economic Literature* 31 (1): 45–104.

Kim, Sukkoo, and Robert A. Margo. 2003. "Historical Perspectives on U.S. Economic Geography." NBER Working Paper 9594. Cambridge, Mass.: National Bureau of Economic Research.

Koehler, Cynthia L. 1995. "Water Rights and the Public Trust Doctrine:

Resolution of the Mono Lake Controversy." *Ecological Law Quarterly* 22: 541–89.

Libecap, Gary D. 1989. *Contracting for Property Rights.* New York: Cambridge University Press.

——— 1998. "Common Property." In *The New Palgrave Dictionary of Economics and the Law,* ed. Peter Newman, Vol. 1: 317–24. London: Macmillan.

——— 2005a. *"Chinatown:* Owens Valley and Western Water Re-allocation— Getting the Record Straight and What It Means for Water Markets." *Texas Law Review* 83 (7): 2055–89.

——— 2005b. *Rescuing Water Markets: Lessons from Owens Valley.* PERC Policy Series, PS-33. Bozeman, Mont.: Property and Environment Research Center.

——— 2006. "Assigning Property Rights in the Common Pool. Implications of the Prevalence of First-Possession Rules." Working paper, Donald Bren School of Environmental Science and Management, University of California, Santa Barbara.

Libecap, Gary D., and James L. Smith. 1999. "The Self-Enforcing Provisions of Oil and Gas Unit Operating Agreements: Theory and Evidence." *Journal of Law, Economics, and Organization* 15 (2): 526–48.

Literary Digest. 1924. "California's Little Civil War." December 6: 13–14.

Loomis, John B. 1987. "Expanding Contingent Value Sample Estimates to Aggregate Benefit Estimates: Current Practices and Proposed Solutions." *Land Economics* 63 (4): 398–402.

Loomis, John B., Katherine Quattlebaum, Thomas C. Brown, and Susan J. Alexander. 2003. *Water Resources Development* 19 (1): 21–28.

Los Angeles Board of Water Commissioners. Various years. *Annual Reports.*

Los Angeles Department of Public Service. 1916. *Complete Report on the Construction of the Los Angeles Aqueduct.* Los Angeles.

Los Angeles Department of Water and Power (LADWP). 1976. "Final Environmental Impact Report on Increased Pumping of the Owens Valley Groundwater Basin." Los Angeles.

——— 2004. "Final Environmental Impact Report, Lower Owens River Project." June 23.

Los Angeles Department of Water and Power (LADWP), Environmental Protection Agency (EPA), and Inyo County. 2002. "Draft Environmental Impact Report/Environmental Impact Statement—Lower Owens River Project." November 1.

Lueck, Dean. 1995. "The Rule of First Possession and the Design of the Law." *Journal of Law and Economics* 38 (2): 393–436.

———— 1998. "First Possession." In *The New Palgrave Dictionary of Economics and the Law*, ed. Peter Newman, Vol. 2: 132–144. London: Macmillan.

Lyon, Robirda. 2002. "The County of Origin Doctrine: Insufficient as a Legal Water Right in California." *San Joaquin Agricultural Law Review* 12: 133–56.

MacDonnell, Lawrence J. 1990. *The Water Transfer Process as a Management Option for Meeting Changing Water Demands*, Vol. 1. Washington, D.C.: USGS.

Meyers, Gary D. 1989. "Variation on a Theme: Expanding the Public Trust Doctrine to Include Protection of Wildlife." *Environmental Law* 19: 723–35.

Miller, Gordon R. 1977. "Los Angeles and the Owens River Aqueduct." PhD diss., Claremont Graduate School, Claremont, Calif.

Mills, Edwin S. 1967. "An Aggregative Model of Resource Allocation in a Metropolitan Area." *American Economic Review* 57: 197–210.

Mulholland, Catherine. 2000. *William Mulholland and the Rise of Los Angeles*. Berkeley: University of California Press.

Nadeau, Remi A. 1950. *The Water Seekers*. Garden City, N.Y.: Doubleday.

National Research Council. 1992. *Water Transfers in the West: Efficiency, Equity, and the Environment*. Committee on Western Water Management, Water Science and Technology Board, Commission on Engineering and Technical Systems, with the Assistance of the Board on Agriculture. Washington, D.C.: National Academy Press.

Nichols, Peter D., and Douglas S. Kenny. 2003. "Watering Growth in Colorado: Swept Along by the Current or Choosing a Better Line?" *University of Denver Water Law Review* 6: 411–52.

Northwest Economic Associates. 2004. "Third-Party Impacts of the Palo Verde Land Management, Crop Rotation and Water Supply Program, Draft Report." Sacramento, Calif., March 29.

Osborne, Henry Z. 1913. "The Completion of the Los Angeles Aqueduct." *Scientific American* 109 (10): 364–71.

Ostrom, Elinor. 1990. *Governing the Commons: The Evolution of Institutions for Collective Action*. Cambridge: Cambridge University Press.

Ostrom, Vincent. 1953. *Water and Politics: A Study of Water Policies and Administration in the Development of Los Angeles*. Los Angeles: Haynes Foundation.

———— 1971. *Institutional Arrangements for Water Resource Development: The Choice of Institutional Arrangements for Water Resource Development*. Arlington, Va.: National Water Commission.

Phillips, William Emerson. 1967. Regional Development of Owens Valley, California: An Economic Base Study of Natural Resources." PhD diss., University of California, Berkeley.

Pisani, Donald J. 1984. *From the Family Farm to Agribusiness: The Irrigation Crusade in California and the West, 1850–1931.* Berkeley: University of California Press.

Provencher, Bill, and Oscar Burt. 1993. "The Externalities Associated with the Common Property Exploitation of Groundwater." *Journal of Environmental Economics and Management* 24: 139–58.

Ralph, Aaron. 2003. "Drain the Water and Pull the Plug on the Economy of One Community So That Another Community Can Brim Over with Economic Development." *McGeorge Law Review* 34: 903–28.

Rauch, James E. 1993. "Productivity Gains from Geographic Concentration of Human Capital: Evidence from the Cities." *Journal of Urban Economics* 34 (3): 380–400.

Reisner, Marc. 1986. *Cadillac Desert: The American West and Its Disappearing Water.* New York: Viking.

Rose, Carol M. 1985. "Possession as the Origin of Property." *University of Chicago Law Review* 52: 73–87.

——— 1998. "Evolution of Property Rights." In *The New Palgrave Dictionary of Economics and the Law,* ed. Peter Newman, Vol. 2: 93–97. London: Macmillan.

Rothman, Hal K. 2002. "The State of the Natural Resources Literature." *Natural Resources Journal* 42: 211–22.

Saliba, Bonnie. 1987. "Do Water Markets 'Work'? Market Transfers and Trade-Offs in the Southwestern States." *Water Resources Research* 23 (7): 1113–22.

Sauder, Robert A. 1994. *The Lost Frontier: Water Diversion in the Growth and Destruction of Owens Valley Agriculture.* Tucson: University of Arizona Press.

Sawyer, Andrew H. 1997. "Changing Landscapes and Evolving Law: Lessons from Mono Lake on Takings and the Public Trust." *Oklahoma Law Review* 50: 311–49.

Sax, Joseph. 1970. "The Public Trust Doctrine in Natural Resource Law: Effective Judicial Intervention." *Michigan Law Review* 68: 471–566.

——— 1990. "The Constitution, Property Rights and the Future of Water Law." *University of Colorado Law Review* 61: 257–82.

——— 1994. "Understanding Transfers: Community Rights and the Privatization of Water." *West-Northwest Journal of Environmental Law and Policy* 1: 13–16.

Schade, Theodore D. 2002. "Testimony Before the California State Water Re-

sources Control Board Hearing Regarding Salton Sea." www.waterrights
.ca.gov/IID/IIDHearingData/LocalPublish/DOW17.pdf.

Segal, Ilya. 1999. "Contracting with Externalities." *Quarterly Journal of Economics* 114 (2): 337–88.

Simms, Richard A. 1995. "A Sketch of the Aimless Jurisprudence of Western Water Law." In *Water Law: Trends, Policies, and Practice*, ed. Kathleen Marion Carr and James D. Crammond, 320–29. Chicago: American Bar Association.

Smith, Rodney T. 2001. *Troubled Waters: Financing Water in the West*. Washington, D.C.: Council of State Planning Agencies.

Starr, Kevin. 2005. *California: A History*. New York: Modern Library.

Stavins, Robert N. 2003. "Market-Based Environmental Policies: What Can We Learn from U.S. Experience (and Related Research)?" Workshop, August 23–24, Donald Bren School of Environmental Science and Management, University of California, Santa Barbara.

Stevens, Jan S. 2003. "The Public Trust and In-stream Uses." *Issues in Legal Scholarship*, Joseph Sax and the Public Trust, Article 9. http://www.bepress.com/ils/iss4/art9.

Tarlock, A. Dan. 2000. "Reconnecting Property Rights to Watersheds." *William and Marry Environmental Law and Policy Review* 25 (Fall): 69–112.

Thompson, Barton H. 1993. "Institutional Perspectives on Water Policy and Markets." *California Law Review* 81: 673–764.

Todd, David. 1992. "Common Resources, Private Rights and Liabilities: A Case Study on Texas Groundwater Law." *Natural Resources Journal* 32 (Spring): 233–63.

U.S. Department of Commerce. Various years. *Census of Agriculture*. Washington, D.C.

Van Valen, Nelson. 1977. "A Neglected Aspect of the Owens River Aqueduct Story: The Inception of the Los Angeles Municipal Electric Story." *Historical Society of Southern California* 59 (Spring): 85–109.

Vorster, Peter. 1992. "The Development and Decline of Agriculture in the Owens Valley." In *The History of Water: Eastern Sierra Nevada, Owens Valley, White-Inyo Mountains*, ed. Clarence A. Hall, Victoria Doyle-Jones, and Barbara Widawski, Vol. 4: 268–84. White Mountain Research Station Symposium. Berkeley: University of California Press.

Walton, John. 1992. *Western Times and Water Wars: State, Culture, and Rebellion in California*. Berkeley: University of California Press.

Wheeler, Mark. 2002. "California Scheming: Los Angeles' Insatiable Thirst for Water, Which Drained the Owens Valley, Has Ruined Lives, Shaped the

City's Politics and Provoked Ongoing Controversy." *Smithsonian* 33 (7): 104–13.

Williamson, Oliver E. 1975. *Markets and Hierarchies: Analysis and Antitrust Implications*. New York: Free Press.

—— 1979. "Transaction-Cost Economics: The Governance of Contractual Relations." *Journal of Law and Economics* 22: 233–61.

Wood, Richard Coke. 1973. *The Owens Valley and the Los Angeles Water Controversy, Owens Valley as I Knew It*. Stockton, Calif.: Pacific Center for Western Historical Studies, University of the Pacific.

Young, Robert A. 1986. "Why Are There So Few Transactions Among Water Users?" *American Journal of Agricultural Economics* (December): 1143–51.

INDEX